D0946417

Treatise on Materials Science and Technology

VOLUME 5

ADVISORY BOARD

G. M. BARTENEV
Academy of Sciences of the USSR
Moscow, USSR

J. W. CHRISTIAN
Oxford University, Oxford, England

M. E. FINE
Northwestern University, Evanston,
Illinois

J. FRIEDEL
Université de Paris, Orsay, France

J. J. HARWOOD
Ford Motor Company, Dearborn,
Michigan

P. B. HIRSCH, F.R.S.
Oxford University, Oxford, England

T. B. KING
Massachusetts Institute of Technology,
Cambridge, Massachusetts

A. SEEGER
Max-Planck-Institut, Stuttgart, Germany

A. SOSIN
University of Utah, Salt Lake City, Utah

F. F. Y. WANG
State University of New York at Stony
Brook, Stony Brook, New York

TREATISE ON MATERIALS SCIENCE AND TECHNOLOGY

EDITED BY

HERBERT HERMAN

Department of Materials Science
State University of New York at Stony Brook
Stony Brook, New York

VOLUME 5

 1974

ACADEMIC PRESS   New York   San Francisco   London

A Subsidiary of Harcourt Brace Jovanovich, Publishers

LIBRARY
University of Texas
At San Antonio

COPYRIGHT © 1974, BY ACADEMIC PRESS, INC.
ALL RIGHTS RESERVED.
NO PART OF THIS PUBLICATION MAY BE REPRODUCED OR
TRANSMITTED IN ANY FORM OR BY ANY MEANS, ELECTRONIC
OR MECHANICAL, INCLUDING PHOTOCOPY, RECORDING, OR ANY
INFORMATION STORAGE AND RETRIEVAL SYSTEM, WITHOUT
PERMISSION IN WRITING FROM THE PUBLISHER.

ACADEMIC PRESS, INC.
111 Fifth Avenue, New York, New York 10003

United Kingdom Edition published by
ACADEMIC PRESS, INC. (LONDON) **LTD.**
24/28 Oval Road, London NW1

LIBRARY OF CONGRESS CATALOG CARD NUMBER: 77-182672

ISBN 0-12-341805-4

PRINTED IN THE UNITED STATES OF AMERICA

Contents

Solution Thermodynamics

Rex B. McLellan

Radiation Studies of Materials Using Color Centers

W. A. Sibley and Derek Pooley

Four Basic Types of Metal Fatigue

W. A. Wood

The Relationship between Atomic Order and the Mechanical Properties of Alloys

M. J. Marcinkowski

List of Contributors

Numbers in parentheses indicate the pages on which the authors' contributions begin.

REX B. McLELLAN (1), Materials Science Department, Rice University, Houston, Texas

M. J. MARCINKOWSKI (181), Engineering Materials Group, and Department of Mechanical Engineering, University of Maryland, College Park, Maryland

DEREK POOLEY (45), Materials Physics Division, Atomic Energy Research Establishment, Harwell, Berkshire, United Kingdom

W. A. SIBLEY (45), Department of Physics, Oklahoma State University, Stillwater, Oklahoma

W. A. WOOD (129), School of Engineering and Applied Science, George Washington University, Washington, D.C.

Preface

Materials limitations are often the major deterrents to the achievement of new technological advances. In modern engineering systems, materials scientists and engineers must continually strive to develop materials which can withstand extreme conditions of environment and maintain their required properties. In the last decade we have seen the emergence of new types of materials, literally designed and processed with a specific use in mind. Many of these materials and the advanced techniques which were developed to produce them, came directly or indirectly from basic scientific research.

Clearly, the relationship between utility and fundamental materials science no longer needs justification. This is exemplified in such areas as composite materials, high-strength alloys, electronic materials, and advanced fabricating and processing techniques. It is this association between the science and technology of materials on which we intend to focus in this treatise.

The topics to be covered in this *Treatise on Materials Science and Technology* will include the fundamental properties and characterization of materials, ranging from simple solids to complex heterophase systems. This treatise is aimed at the professional scientist and engineer, as well as the graduate students in materials science and associated fields.

This fifth volume of the *Treatise* covers the areas of thermodynamics, radiation effects, and mechanical properties. The four articles in this volume are strongly tutorial, and we hope to repeat this approach in future volumes.

In the first article, McLellan reviews solution thermodynamics, from the basic equations and concepts through modern research problems. Sibley and Pooley, in their contribution, combine a fundamental and applied approach to the study of radiation-induced color centers. There are two articles on mechanical properties: Wood examines fatigue with a strong emphasis on metallography, and Marcinkowski writes on strength and atomic order, an area of research long in need of a thorough review.

Forthcoming volumes will continue to examine a wide range of topics in the fields of metals, ceramics, and polymers. In addition, there will be volumes devoted to specific topics, as well as monographs on particularly active subjects.

The editor would once again like to express his sincere appreciation to the members of the Editorial Advisory Board who have given so generously of their time and advice.

H. HERMAN

Contents of Previous Volumes

Solution Thermodynamics

REX B. McLELLAN

Materials Science Department
Rice University
Houston, Texas

I. Introduction

The large increase seen in recent years in the output of papers in the general field of solution thermodynamics indicates that a review of this topic is appropriate at this time. The number of subtopics that may properly be included under the general heading of solution thermodynamics is large. Investigations ranging from the computer simulation calculations of point defect aggregates to diffusion kinetics may be considered to fall into this category, so that a review of solution thermodynamics in its widest sense would be an undertaking of major proportions. Accordingly the present review will attempt to cover certain topics with reasonable thoroughness, whereas others will be omitted. In a recent short review (McLellan and Chraska, 1971) the thermodynamics of the technologically important iron-based solutions containing carbon have been discussed. In a second recent review (McLellan, 1972) statistical mechanical models for solid solutions were emphasized.

The thrust of the current review will be more general so as to appeal to a wider pedagogical base, and an attempt will be made to include some topics that were omitted in the previous two papers. Specifically, sections dealing with hydrogen–metal solutions, liquid metal solutions, and the relationship between constant-pressure and constant-volume ensembles will be included in the review. These three areas, particularly the subject of hydrogen–metal systems, have been the object of a recent upsurge of interest.

, With regard to the use of units and symbols, the present review will adhere to the normal practice in the United States of employing the calorie for energies rather than joules. There is currently much confusion regarding the symbols to be used for the thermodynamic potentials and their derivatives with respect to concentration units. In this respect we will again use the American norm of denoting the Helmholtz free energy by F and the Gibbs free energy by G. For the solute concentration in substitutional solutions, the atom fraction of solute i will be denoted by c_i. If the solution has two components, the major component (solvent) will be denoted by v and the minor component (solute) by u. In the case of binary interstitial solutions the appropriate concentration unit for most purposes is the atom ratio. This is the ratio of the number of gram atoms of solute to solvent. Atom ratios will be denoted by θ. In many cases, especially in liquid solutions, the structural information enabling a choice between *interstitial* or *substitutional* to be made is lacking (or neither term is appropriate anyway). In such cases the atom fraction will be used as the concentration unit.

It should also be pointed out that some degree of overlap between the material presented in this review and that discussed in the two previous reviews mentioned here will be inevitable. Much repetition will be avoided, but some will be necessary in order to preserve continuity and widen the pedagogical appeal of the present review.

II. Formal Relations for Solutions

A. Partial Molar Quantities

The most important thermodynamic variables used in discussing the properties of solutions are the partial molar quantities. If we imagine that a number of moles dn_1 of component 1 is added reversibly to an infinitely large volume of solution at constant pressure and temperature and without changing the number of moles of the other components, and the resulting change in the extensive thermodynamic function Q is dQ, then the partial molar quantity of component 1 in solution is

$$\bar{Q}_1 = (\partial Q/\partial n_1)_{P, T, n_2, n_3, \ldots} \tag{1}$$

These quantities can be defined for all the extensive state functions of the solution and are subject to interrelations of the same form as those relating the concomitant integral thermodynamic quantities. If S denotes the entropy, E the internal energy, and V the volume, then

$$\bar{V}_1 = (\partial \bar{G}_1/\partial P)_{T, n_j} \tag{2}$$

$$\bar{S}_1 = -(\partial \bar{F}_1/\partial T)_{V, n_j} = -(\partial \bar{G}_1/\partial T)_{P, n_j} \tag{3}$$

$$\bar{H}_1 = \bar{E}_1 + P\bar{V}_1 \tag{4}$$

$$\bar{F}_1 = \bar{E}_1 - T\bar{S}_1 \tag{5}$$

$$\bar{G}_1 = \bar{H}_1 - T\bar{S}_1 \tag{6}$$

Of all the partial molar quantities, the most important is the partial molar Gibbs free energy. This quantity, or simple translations of it, is the quantity that has found the widest usage in discussing the thermodynamic properties of solutions. It can be written in several ways:

$$\bar{G}_1 \equiv \mu_1 = (\partial G/\partial n_1)_{P, T, n_j} = (\partial H/\partial n_1)_{P, S, n_j} = (\partial F/\partial n_1)_{V, T, n_j} \tag{7}$$

In addition to the relations outlined above it is easy to show that the molar value Q_m of an extensive quantity Q is given by

$$Q_m = c_1 \bar{Q}_1 + c_2 \bar{Q}_2 + \cdots \tag{8}$$

and, from this that, at constant pressure and temperature,

$$c_1 d\bar{Q}_1 + c_2 d\bar{Q}_2 + \cdots = 0 \tag{9}$$

If Q denotes the Gibbs free energy of a binary solution, Eq. (9) is known as the Gibbs–Duhem relation. An excellent discussion of Eqs. (8) and (9) has been given by Darken and Gurry (1953). Another relation of great importance can be derived from Eqs. (8) and (9). For a binary system, this is

$$\bar{Q}_u = Q_m + (1 - c_u)\, dQ_m/dc_u, \qquad \bar{Q}_v = Q_m + (1 - c_v)\, dQ_m/dc_v \quad (10)$$

where the v (solvent) and u (solute) nomenclature has been used. The relations (10) form the basis of a method of obtaining values of $\bar{Q}_u$ and $\bar{Q}_v$ from experimental data for the variation of Q_m with composition. This is illustrated in Fig. 1. The curved line in Fig. 1 represents the experimental

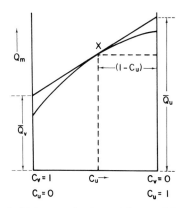

Fig. 1. Method of intercepts for determining partial molar quantities.

data for Q_m. At the point X a tangent is constructed. From Eqs. (10) it can be seen by inspection that its ordinate at $c_u = 0$ gives $\bar{Q}_v$ for the composition X and the ordinate at $c_v = 0$ gives $\bar{Q}_u$. It is also clear from the diagram that the partial molar quantities are not known absolutely and must be calculated with respect to the same reference state used for Q. This is not true, however, for the relative partial molar quantities

$$\Delta\bar{Q}_i = \bar{Q}_i - Q_i^\circ \tag{11}$$

where Q_i° is the value of Q for the pure component i per mole under given standard conditions of the relevant intensive variables.

The usefulness of the relative partial molar quantities in comparing the properties of a series of solutes in a given solvent metal is limited to some extent by the fact that Q_i°, which is not relevant to the physics of the solution per se, may be much larger than $\bar{Q}_i$. An example of this occurs in writing $\Delta\bar{S}_i$ for solutions when the pure solute in the standard state is gaseous. This leads to large negative values for $\Delta\bar{S}_i$. Thus it is often advantageous, when comparing the properties of solutes in a given solvent, to use the partial molar quantity $\bar{Q}_i$, relative to some physically appropriate reference state such as a solute atom at rest in a vacuum.

B. The Ideal Solution

There are several alternative procedures used to discuss the thermodynamic behavior of real solutions. Many of these procedures utilize a comparison with the ideal solution, in a similar way that real gases are compared with ideal gases.

A solution may be termed ideal if the relative partial volumes and enthalpies of its components are zero and the entropy of the solution arises only from the completely random mixing of the components. For a binary system, the first two conditions can be written

$$\Delta \bar{V}_u^{id} = 0 \tag{12}$$

$$\Delta \bar{H}_u^{id} = 0 \tag{13}$$

but in writing the third condition in an explicit form, generality is lost; the form of $\Delta \bar{S}_u^{id}$ is determined by the structure of the solution. For a substitutional solution,

$$\Delta \bar{S}_u^{id} = -k \ln c_u \tag{14}$$

and for an interstitial solution in which there are β sites per solvent (v) atom,

$$\Delta \bar{S}_u^{id} = -k \ln \frac{\theta_u/\beta}{1 - (\theta_u/\beta)} \tag{15}$$

The corresponding chemical potentials are given by

$$\mu_u^{id} = \mu_u^\circ + kT \ln c_u \tag{16}$$

and

$$\mu_u^{id} = \mu_u^\circ + kT \ln \frac{\theta_u/\beta}{1 - (\theta_u/\beta)} \tag{17}$$

The quantity μ_u° is $H_u^\circ - TS_u^\circ$. There are no contributions to the free energy arising from nonconfigurational effects. Thus, the ideal solution does not exist, although a solution of isotopes of the same element would approach ideality closely.

Solutions in which the entropy of position is ideal, although the other thermodynamic functions are nonideal, are termed regular.

C. Activities and Activity Coefficients

The activity α_u of a solute species, with respect to the pure solute, is normally defined in analogy with Eq. (16) in the form

$$\mu_u = \mu_u^\circ + kT \ln \alpha_u \tag{18}$$

and the thermodynamic properties are described by the variation of α_u with composition. If $\alpha_u = c_u$ the solute is said to obey Raoult's law; and if, over a limited composition range α_u is directly proportional to c_u, the solute obeys Henry's law in that composition range. It must be remembered that while the definition (18) is appropriate for substitutional solutions, the curve $\alpha_u = c_u$ does not imply random mixing in the interstitial case. For the interstitial solution where, for simplicity, $\beta = 1$,

$$\alpha_u = \frac{\theta_u}{1 - \theta_u} = \frac{c_u}{1 - 2c_u} \tag{19}$$

implies ideal behavior (see the upper solid curve in Fig. 2). Figure 2 also shows, in dashed lines, $\Delta \bar{S}_u^{id}$ for the substitutional and interstitial cases.

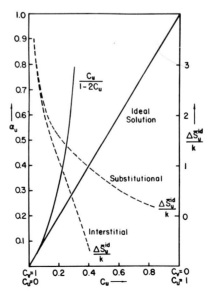

Fig. 2. The relative partial molar configurational entropy for ideal substitutional and interstitial solutions. The solid lines give the solute activities.

An alternative representation of solute behavior is based on the activity coefficient γ_u, defined by

$$\gamma_u = \alpha_u/c_u \tag{20}$$

so that

$$\mu_u = \mu_u^\circ + kT \ln c_u + kT \ln \gamma_u$$
$$= \mu_u^{id} + kT \ln \gamma_u \tag{21}$$

and $\gamma_u = 1$ for the ideal solution.

Yet another representation of the thermodynamic behavior of solute atoms in solution was introduced by Scatchard (1949), who defined the excess partial molar quantities in the form

$$\bar{Q}_u^{xs} = \bar{Q}_u - \bar{Q}_u^{id} \tag{22}$$

so that the excess partial molar Gibbs free energy of a substitutional solute is

$$\mu_u^{xs} = \mu_u - \mu_u^{id} = \mu_u - \mu_u^\circ - kT \ln c_u$$
$$= kT \ln \gamma_u \tag{23}$$

The quantities α_u, γ_u, and $\bar{Q}_u^{xs}$ are all useful means of describing the thermodynamic behavior of solutions and they are used with about equal regularity. Recently the Wagner interaction coefficients have gained popularity as a means of discussing solute behavior in solutions. The first-order free energy interaction coefficient $\varepsilon_i^{(j)}$ of component i in a multicomponent dilute solution was defined by Wagner (1962) in the form

$$\varepsilon_i^{(j)} = \lim_{c_1 \to 1} (\partial \ln \gamma_i/\partial c_j) \tag{24}$$

where c_1 denotes the atom fraction of the solvent and γ_i is the activity coefficient of the component i. This interaction coefficient is the first-order term in a Taylor series expansion of the partial excess free energy of i

$$\ln \gamma_i = \ln \gamma_i^\circ + \sum_{i \neq j} \varepsilon_i^{(j)} c_j + \text{higher terms} \tag{25}$$

The concept of interaction coefficients has been generalized by Lupis and Elliott (1965) to include similar coefficients for the excess partial entropy and enthalpy and also higher terms in the expansion (Lupis and Elliott, 1966a,b). It has, however, been pointed out by Parris and McLellan (1972) that experimentally determined $\varepsilon_i^{(i)}$ values do not allow a distinction to be made between physically disparate models which are equally compatible with measured activity data and, furthermore, the $\varepsilon_i^{(i)}$ values determined from measurements in the composition range where solute atom mutual interactions lead to deviations from nonrandom mixing behavior need not be compatible with $\varepsilon_i^{(i)}$ values calculated from valid models in the limit when c_i approaches zero. This is true despite the fact that $\varepsilon_i^{(i)}$ is defined in this limit.

Recently various formalisms have been proposed for the correlation of sets of thermodynamic data. Darken (1967) has pointed out that for many binary liquid solutions the two terminal regions can often be described by writing the activity coefficient of the solvent in the form

$$\ln \gamma_v = \alpha^1 (1 - c_v)^2 \tag{26}$$

and that of the solute in the form

$$\ln \gamma_u = \alpha^1 (1 - c_u)^2 + I \tag{27}$$

where α^1 and I are constants. This quadratic formalism has been extended by Turkdogan and Darken (1968) to include expressions for the enthalpy and entropy of mixing and the data for many liquid binary solutions have been shown by Turkdogan *et al.* (1969) to be consistent with the quadratic formalism.

D. Integration of the Gibbs–Duhem Equation

If the partial molar quantity of one component in a binary solution is known, the other may be found from an integration of the Gibbs–Duhem equation (9). It is usual to integrate from $c_u = 1$ to c_u, so that

$$\bar{G}_u - G_u^\circ = -\int_{c_u=1}^{c_u} (c_v/c_u) \, d\bar{G}_v \tag{28}$$

As c_u approaches zero, the integrand approaches infinity so that difficulties arise in performing the integration in this region. These difficulties can be overcome by the employment of alternative variables. Darken and Gurry (1953) show that if the variable

$$a_i \equiv (\ln \gamma_i)/(1 - c_i)^2 \tag{29}$$

is used, the integrated Gibbs–Duhem equation for binary systems can be put into the form

$$\ln \gamma_u = -a_v c_u c_v - \int_{c_u=1}^{c_u} a_v \, dc_u \tag{30}$$

The function a_i is always finite since solutions obey Raoult's law in the composition region where c_i approaches unity. Using this function allows the graphical integration to be performed without difficulty.

Darken (1950) extended the application of the integrated Gibbs–Duhem equation to ternary solutions. Note that Eqs. (10) can be generalized in the form

$$\bar{G}_2 = G_m + (1 - c_2)(\partial G_m/\partial c_2)_{c_1/c_3} \tag{31}$$

where c_1, c_2, and c_3 denote the atom fractions in a ternary solution and the derivative is evaluated holding the proportions of components 1 and 3 constant. Attention is thus being focused on the variation of $\bar{G}_2$ along a composition line such as 2-A in Fig. 3. Dividing Eq. (31) by $(1 - c_2)^2$ and integrating from $c_2 = 1$ to c_2 gives

$$G_m - (1 - c_2) \lim_{c_2 \to 1} \left\{ \frac{G_m}{1 - c_2} \right\} = (1 - c_2) \int_1^{c_2} \frac{\bar{G}_2}{(1 - c_2)^2} \, dc_2 \qquad (32)$$

where both the limit and the integral are taken at constant c_1/c_3. Writing

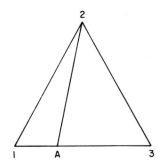

Fig. 3. Composition variation of $\bar{G}_2$.

Eq. (32) in terms of the excess molar Gibbs free energy G_m^{xs} and noting the compositional relationships

$$c_1 = \frac{1 - c_2}{1 + (c_3/c_1)} \; ; \qquad c_3 = \frac{1 - c_2}{1 + (c_1/c_3)} \qquad (33)$$

it can be shown that,

$$G_m^{xs} = (1 - c_2) \int_1^{c_2} \frac{\bar{G}_2^{xs}}{(1 - c_2)^2} \, dc_2 + c_1 [\bar{G}_1^{xs}]_{c_2=1} + c_3 [\bar{G}_3^{xs}]_{c_2=1} \qquad (34)$$

where the two terms $[\bar{G}_1^{xs}]_{c_2=1}$ and $[\bar{G}_3^{xs}]_{c_2=1}$ are constants determined from measurements on the 1–2 and 2–3 binary systems. Regarding $\bar{G}_2^{xs}$ in the ternary solution as being the experimentally determined quantity, these two constants can be determined by applying the Gibbs–Duhem relation to the two binary systems

$$[\bar{G}_1^{xs}]_{c_2=1} = -\left\{ \int_1^0 \frac{\bar{G}_2^{xs}}{(1 - c_2)^2} \, dc_2 \right\}_{c_3=0} \qquad (35)$$

$$[\bar{G}_3^{xs}]_{c_2=1} = -\left\{ \int_1^0 \frac{\bar{G}_2^{xs}}{(1 - c_2)^2} \, dc_2 \right\}_{c_1=0} \qquad (36)$$

Equations (34)–(36) express the excess molar free energy of a ternary solution in terms of the partial molar free energy of only one component in the solution. Wagner (1962) has given an alternative treatment of the ternary Gibbs–Duhem problem in which different concentration units are used.

Darken (1950) also showed that the excess free energy of a binary system (say 1–3) can be found from measurements of $\bar{G}_2^{xs}$ for component 2 in the ternary solution. Since the binary 1–3 system is the limiting case of the 1–2–3 ternary system when $c_2 \to 0$ we can find G_{1-3}^{xs} by setting $c_2 = 0$ in Eq. (34). This gives

$$G_{1-3}^{xs} = \left\{ \int_1^0 \frac{\bar{G}_2^{xs}}{(1-c_2)^2} \, dc_2 \right\}_{c_1/c_3} - c_1 \left\{ \int_1^0 \frac{\bar{G}_2^{xs}}{(1-c_2)^2} \, dc_2 \right\}_{c_3=0}$$
$$- c_3 \left\{ \int_1^0 \frac{\bar{G}_2^{xs}}{(1-c_2)^2} \, dc_2 \right\}_{c_1=0} \tag{37}$$

If the approximation is made that $\bar{G}_2^{xs}/(1-c_2)^2$ is a function of the ratio c_1/c_3 only, Eq. (37) yields, since $\bar{G}_2^{xs} = 0$ when $c_2 = 1$,

$$G_{1-3}^{xs} = -\bar{G}_2^{xs}(c_2 = 0) + c_1 \bar{G}_2^{xs}(c_1 = 1) + c_3 \bar{G}_2^{xs}(c_3 = 1) \tag{38}$$

Rewriting this in terms of the activity coefficients from Eq. (23) we obtain

$$\ln \gamma_2^{1-2-3} = c_1 \ln \gamma_2^{1-2} + c_3 \ln \gamma_2^{2-3} - G_{1-3}^{xs}/kT \tag{39}$$

for dilute solutions of component 2 dissolved in a 1–3 solvent. The assumption that $\bar{G}_2^{xs}/(1-c_2)^2$ is dependent on only the ratio c_1/c_3 is equivalent to assuming regular behavior in the ternary system. Alcock and Richardson (1958) derived the same equation using a zeroth-order model.

Now if component 2 is an interstitial solute, problems arise since the integrals cannot be evaluated. However, Oates and McLellan (1968) showed that if 2 was an interstitial component and the ternary solution was randomly mixed,

$$\ln \gamma_2^{1-2-3} = c_1 \ln \gamma_2^{1-2} + c_3 \ln \gamma_2^{2-3} \tag{40}$$

and the term involving G_{1-3}^{xs} does not appear. In such solutions the thermodynamic properties of the dissolved interstitial solute are independent of the properties of the solvent. Wada and Saito (1961) have demonstrated that in such solutions the Wagner interaction coefficients are again independent of the solvent atom interactions, whereas $\varepsilon_i^{(i)}$ for ternary solutions in which all three components are dissolved substitutionally does contain a term involving interactions between the atoms in the 1–3 solvent lattice. The literature contains many examples of the application of Eq. (39) to solutes in ternary systems where the solute in question is dissolved interstitially.

E. Temperature Dependence of $\Delta \bar{G}_i$

In a given solution $\Delta \bar{G}_i$ and $\bar{G}_i$ can vary with temperature and composition. In more concentrated solutions where solute atom mutual interactions play a large part in determining the thermodynamic properties, $\Delta \bar{G}_i$ can often vary with temperature, at a fixed composition, through the temperature variation of $\Delta \bar{H}_i$ and $\Delta \bar{S}_i$.

However, in the infinitely dilute solution where the positional entropy can be expected to be ideal and independent of temperature, the temperature variation of $\Delta \bar{H}_i$ and $\Delta \bar{S}_i$ may be principally due to the standard state functions H_i° and S_i° and not to the physically important quantities $\bar{H}_i$ and $\bar{S}_i$. This can be illustrated by the very extensive thermodynamic data on solid solutions of carbon in fcc iron given by Ban-ya *et al.* (1969, 1970). Their data covered composition ranges from very dilute solutions up to the saturation solubility at temperatures in the range 900–1400°C. Figure 4 shows a plot of $\Delta \bar{G}_u^\infty$, the value of $\Delta \bar{G}_u$ in the infinitely dilute solution, as a function of temperature. The plot is distinctly curved. However, the plot of

$$\Delta \bar{G}_u^\infty + G_{gr}^\circ = \bar{H}_u^\infty - T\bar{S}_u^\infty = \bar{G}_u^\infty \tag{41}$$

against temperature is linear. A least-squares regression of these data shows that, from the slope and intercept, $\bar{H}_u^\infty = -155.28$ kcal/mole and $\bar{S}_u^\infty = 5.78$ k. The standard state is an atom of carbon at rest in a vacuum. The temperature variation of G_{gr}° is also shown in Fig. 4. These remarks do

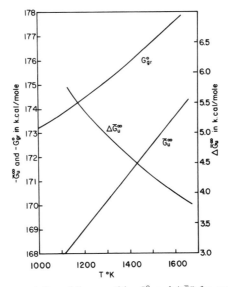

Fig. 4. Temperature variation of the quantities G_{gr}^0 and $\Delta \bar{G}_u^\alpha$ for carbon in fcc iron.

not prove conclusively that $\bar{H}_u^\infty$ and $\bar{S}_u^\infty$ for carbon in austenite are independent of temperature, but they provide strong evidence for this. The measurements of the composition and temperature dependence of the free energy of carbon in bcc iron in the temperature range 575–825°C (Dunn and McLellan, 1971) also showed that $\bar{H}_u^\infty$ and $\bar{S}_u^\infty$ were independent of temperature to within very close limits and $\bar{H}_u^\infty = -144.06$ kcal/mole and $\bar{S}_u^\infty = 6.56$ k.

In cases where $\Delta\bar{G}_i$ is measured with respect to a gaseous standard state, H_i° and S_i° can be even more temperature dependent. Thus for H_2 molecules for example $H_T^\circ - H_0^\circ$ (the subscript refers to the absolute temperature) varies from 3.429 kcal/mole at 500°K to 6.964 kcal/mole at 1000°K.

Thus, it is suggested that wherever the experimental data are sufficiently extensive and accurate, the use of a temperature-independent standard state is to be recommended.

III. Phase Relations

The study of equilibrium relations between solutions and other phases is one of prime importance to solution thermodynamics. The most important general method of obtaining thermodynamic data on solutions is to investigate the temperature dependence of the equilibrium between the solution and a phase whose thermodynamic properties are known.

A. Solid–Liquid Equilibrium

Wagner (1954) has given a comprehensive account of the equilibrium between a liquid and solid binary solution where both components are completely miscible in both states of aggregation.

Starting from the fact that the chemical potentials of each component are equal when the solid and liquid are in equilibrium, Wagner has calculated the relation between the terminal slopes of the liquidus and solidus

$$(\partial T/\partial c_u^{(l)})_{c_u^{(l)}=0} \equiv K^l \quad \text{and} \quad (\partial T/\partial c_u^{(s)})_{c_u^{(s)}=0} \equiv K^s$$

and the difference between the molar excess free energy $G_m^{xs(s)} - G_m^{xs(l)} \equiv \delta G^{xs}$ in the form

$$\left(\frac{\partial \delta G^{xs}}{\partial c_u}\right)_{c_u=0} = -(\delta T^m)\,\Delta S_u^\circ - RT_v^m \ln\left(1 + \frac{\Delta S_u^\circ}{RT_v}K^l\right) \tag{42}$$

and

$$\left(\frac{\partial \delta G^{xs}}{\partial c_u}\right)_{c_u=0} = -(\delta T^m)\,\Delta S_u^\circ + RT_v^m \ln\left(1 - \frac{\Delta S_u^\circ}{RT_v}K^s\right) \tag{43}$$

where the superscripts l and s denote liquid and solid, T_v^m is the melting temperature of pure solvent, T_u^m is the melting temperature of pure solute, and $\delta T^m = T_v^m - T_u^m$, and ΔS_u° is the entropy of fusion of solute.

Wagner has applied these equations to the calculation of δG^{xs} for Au–Cu solutions at a composition of 40 at. % Cu. This result, $\delta G^{xs} = 418$ cal/mole, is in excellent agreement with that obtained by Wagner from the minimum of the liquidus and solidus lines in the Au–Cu system ($\delta G^{xs} = 421$ cal/mole).

It should be noted that for many binary systems the gap between liquidus and solidus is small and the estimation of K^l and K^s from experimental data is prone to considerable error.

Scheil (1943) and Chipman (1948) have studied the liquidus–solidus relations when the solid phase is virtually pure solvent and the liquid is a regular solution. This leads to simple relations by which the activities in the liquid can be related to the phase diagram. Kubaschewski *et al.* (1967) show that the solvent activity α_T in the liquid at some temperature T is given by

$$\ln \alpha_T = -\frac{(T_v^m - T_c)L_f}{RTT_v^m} + \frac{T - T_c}{T} \ln c_v \tag{44}$$

where T is an arbitrary temperature at which the activity is required, T_c is the liquidus temperature corresponding to the composition c_v of the liquid, and L_f is the heat of fusion of pure solvent.

However, the uncertainties likely to be found in phase equilibrium data coupled with the simplicity of the models assumed for the liquid and solid render the use of equations like (44) doubtful.

B. Equilibrium with a Gas

Gas–solid or gas–liquid equilibrium investigations form an important tool in the study of solutions since the thermodynamic properties of simple gases or gas mixtures of a given temperature and pressure are easy to calculate from statistical mechanics.

The simplest case is where the solution is in equilibrium with a monatomic gas of solute atoms at a pressure P and temperature T. Since, in equilibrium, the chemical potentials of u atoms in the gas and condensed phase are equal, we can equate the chemical potential of u atoms in the gas,

$$\mu_u^g = kT \ln \frac{h^3}{(2\pi mkT)^{3/2}\omega} + kT \ln \frac{P}{kT} \tag{45}$$

(e.g., see Fowler and Guggenheim, 1949) with that in the solid

$$\mu_u^s = H_u^\circ - TS_u^\circ + kT \ln \alpha_u \tag{46}$$

to obtain

$$\alpha_u = \frac{h^3 P}{(2\pi m)^{3/2}(kT)^{5/2}\omega} \exp\left(-\frac{H_u^\circ}{kT}\right) \exp\left(\frac{S_u^\circ}{k}\right) \tag{47}$$

where ω is the electronic ground state degeneracy of the atoms in the gas phase and m is the mass of a u atom. A classical example of the use of this relation has been given by Johnson and Shuttleworth (1959) in their study of the solubility of krypton in liquid lead, tin, and silver. Using fission-product ^{84}Kr they were able to measure solute concentrations as low as $c_u \sim 10^{-11}$ when P was one atmosphere.

If on the other hand the solution is in equilibrium with a diatomic gas, we have in the gas phase

$$\mu_u^g = -\frac{E_D^0}{2} + kT \ln\left|\frac{h^3}{(2\pi 2mkT)^{3/2}}\frac{\sigma h^2}{8\pi^2 IkT}\right|^{1/2}\frac{1}{(kT)^{1/2}} + kT \ln P_{u_2}^{1/2} \tag{48}$$

where E_D^0 is the dissociation energy of the u_2 molecule at $0°$K, I its moment of inertia, and σ its symmetry number. The electronic terms have been omitted for simplicity. Thus, in equilibrium, the solute activity in equilibrium with the u_2 gas at a pressure P_{u_2} and temperature T is given by

$$\alpha_u = \frac{\phi P_{u_2}^{1/2}}{T^{7/4}} \exp\left|-\frac{1}{kT}\left(H_u^\circ + \frac{E_D^0}{2}\right)\right| \exp\left(\frac{S_u^\circ}{k}\right) \tag{49}$$

where

$$\phi = \left|\frac{h^3}{(2\pi 2m)^{3/2}}\frac{\sigma h^2}{8\pi^2 I}\right|^{1/2}\frac{1}{(k)^{7/4}} \tag{50}$$

These equations have been used in analyzing studies of the equilibrium between S_2 molecules and copper–sulfur solid solutions (Moya *et al.*, 1971), between O_2 molecules and Cu–O solutions (Pastorek and Rapp, 1969), between O_2 molecules and Ag–O solutions (McLellan, 1964), and between N_2 molecules and Fe–N solutions (McLellan, 1964).

For equilibrium solubility experiments where the condensed phase is equilibrated with a mixture of gases, similar but rather more complex relations for α_u can be written. They have been used extensively, particularly in studying equilibrium data between iron–carbon solutions and mixtures of $H_2 + CH_4$ or $CO + CO_2$ (see McLellan, 1965; Dunn and McLellan, 1971).

An alternative procedure is to use tabulated free energy data for gases. However, the use of such closed equations for α_u represented by Eqs. (47) and (49) takes the temperature dependence of the free energy of the gas molecules into account, and furthermore the introduction of suitable theoretical expressions for α_u into these equations leads to closed solubility equations from which $\bar{H}_u$ and $\bar{S}_u$ can be derived with respect to a temperature-independent reference level.

It should also be noted that studies of metastable gas–metal equilibria can be useful. For example, at one atmosphere pressure, the solubility of N in austenite in equilibrium with N_2 molecules has a maximum of only 0.03 wt. %. However, in the recent investigation of the metastable NH_4–Fe equilibrium Atkinson and Bodsworth (1970) showed that much more N could be dissolved in the metal lattice.

Another useful technique involving gas–metal equilibria is the vapor transport method in which the solution under investigation is allowed to achieve equilibrium with a gas phase. The gas is also in contact and in equilibrium with a second condensed phase of known thermodynamic properties. The composition and pressure of the gas need not be known. The solute atoms in the gas have the same chemical potential as those in the solution and in the known reference phase. This technique has been used to study the thermodynamic properties of carbon in ternary solid solutions by Zupp and Stevenson (1966) and Chraska and McLellan (1970) and in binary solutions by Dunn et al. (1968), Siller et al. (1968), and McLellan (1969).

An excellent general discussion of experimental methods involving the equilibration of gaseous and condensed phases has been given by Kubaschewski et al. (1967).

C. Solid–Solid Equilibrium

Consider the solvus line in the simple binary phase equilibrium diagram in Fig. 5. At some temperature T the tie-line connects the two terminal solid solutions in equilibrium. If we assume that the dilute solution of u in v is regular, then

$$\mu_u = \bar{H}_u - T\bar{S}_u^{xs} + kT \ln c_u \tag{51}$$

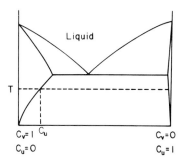

Fig. 5. Equilibrium between two terminal solid solutions in a binary eutectic system.

If this solution is in equilibrium with essentially pure solute u,

$$\mu_u = H_u^\circ - TS_u^\circ$$

and

$$c_u = \exp - \left(\frac{\Delta \bar{H}_u}{kT}\right) \exp \left(\frac{\Delta \bar{S}_u^{xs}}{k}\right) \tag{52}$$

Thus, for the appropriate kind of binary system, the relative partial molar enthalpy and the partial molar excess entropy can be determined from the solvus line.

Freedman and Nowick (1958) have extracted such data from a series of binary equilibrium diagrams. It should, however, be remembered that the reliability of solid–solid equilibrium data, particularly at low temperatures, is often questionable and if the entropy of position in the dilute solution is not truly ideal, the analysis is no longer valid.

IV. Models for Solid Solutions

A. Infinitely Dilute Solutions

In the infinitely dilute solution, the solute concentration is so small that the properties of the solution are not determined by interactions between the solute atoms. The partial enthalpy $\bar{H}_u^\infty$ and excess entropy $\bar{S}_u^{xs}$ are important quantities because they relate directly to theoretical models involving solute–solvent interactions without the encumberance of solute–solute interactions. The quantity $\bar{S}_u^{xs}$ in the infinitely dilute solution will be designated $\bar{S}_u^v$. The reason for this is that, since u–u interactions can be ignored, the configurational entropy is ideal and $\bar{S}_u^{xs}$ arises only from nonconfigurational effects such as changes in the vibrational modes of the solvent lattice on introducing a solute atom. No attempt will be given here to discuss the calculations of $\bar{H}_u^\infty$ and $\bar{S}_u^v$ which range from linear elastic approximations to refined quantum-mechanical treatments. The simplest statistical model whose properties correspond to those of the infinitely dilute solution is the quasi-regular model in which $\bar{H}_u$ is independent of composition and $\bar{S}_u^{xs} = \bar{S}_u^v$ and is also independent of composition. We can write the canonical partition function $Q(v, T)$ of the quasi-regular solution in the form

$$Q(v, T) = Q^i(v, T) \sum_q g(E_q) \exp(-E_q/kT) \tag{53}$$

taking the sum over the accessible quantum states of energy E_q and degeneracy $g(E_q)$. The factor $Q^i(v, T)$ accounts for the internal degrees of freedom. In the quasi-regular solution containing N_u solute and N_v solvent atoms, $Q^i(v, T)$ is given by

$$Q^i(v, T) = \exp\left((N_u \bar{S}_u^v + N_v S_v^\circ)/k\right) \tag{54}$$

and the summation in (53) is replaced by the single degenerate energy level

$$E_q = N_u \bar{E}_u^\infty + N_v E_v^\circ \tag{55}$$

In the substitutional case $g(E_q)$ is simply the number of ways of arranging the N_u mutually indistinguishable solute atoms and N_v mutually indistinguishable solvent atoms on $N_u + N_v$ labeled sites

$$g(E_q) = (N_v + N_u)!/N_v! N_u! \tag{56}$$

Thus

$$Q(v, T) = \frac{(N_v + N_u)!}{N_v! N_u!} \exp\left(\frac{N_u \bar{S}_u^v + N_v S_v^\circ}{k}\right) \exp\left(-\frac{1}{kT}(N_u \bar{E}_u^\infty + N_v E_v^\circ)\right) \tag{57}$$

The chemical potential of the solute is found from

$$F_u = -kT \ln Q(v, T) \tag{58}$$

$$\mu_u = (\partial F_u/\partial N_u)_{T, V, N_v} \tag{59}$$

in the form

$$\mu_u = \bar{E}_u^\infty - T\bar{S}_u^v + kT \ln c_u \tag{60}$$

Note that μ_u has been obtained by differentiating the Helmholtz free energy with respect to N_u at constant volume. This has lead to an equation for μ_u containing $\bar{E}_u^\infty$ not $\bar{H}_u^\infty$. This point is not too important in discussing infinitely dilute solutions but it is important in discussing more-concentrated solutions. This point will be developed in greater detail later.

The quasi-regular model has found considerable application to dilute solid solutions. For many interstitial solutions, the thermodynamic properties have been measured at very small solute concentrations. In such cases the solute chemical potential can be written

$$\mu_u = \bar{E}_u^\infty - T\bar{S}_u^v + kT \ln \frac{\theta_u/\beta}{1 - (\theta_u/\beta)} \tag{61}$$

A compilation of values of $\bar{E}_u^\infty$ and $\bar{S}_u^v$ found for dilute interstitial solid solutions containing C and N is given in Table I. The energy zero for $\bar{E}_u^\infty$ is that of an atom at rest in a vacuum.

TABLE I

Solvent v	Solute u	$\bar{E}_u^\infty$ (kcal/mole)	$\dfrac{\bar{S}_u^v}{k}$
bcc Fe	C	-146.4	6.22
fcc Fe	C	-155.2	5.77
Ni	C	-155.1	3.95
Co	C	-150.5	6.06
Cu	C	-155.47	0.07
Ag	C	-152.95	1.57
Au	C	-147.71	3.20
bcc Fe	N	-74.3	7.20
fcc Fe	N	-81.9	6.70

B. The Zeroth Approximation

As the concentration of solute atoms is increased, u–u interactions become important. The simplest statistical model that attempts to take this into account is the zeroth approximation in which the energy of the solution is written in terms of pairwise interaction energies that are independent of composition (and volume). If we consider a substitutional binary solution containing N_u solute and N_v solvent atoms distributed in a lattice of coordination number z, three kinds of nearest neighbor pairs will occur. These are u–u, v–v, and u–v pairs. If the number of u–v pairs is denoted by λ, the pairs can be enumerated and assigned energies according to Table II.

The energy assignments are based on writing that of a u–u pair as $-2X_u/z$ and that of a u–v pair as

$$\varepsilon_{uv} = (-X_v - X_u + \omega)/z \tag{62}$$

which defines the interchange energy ω. The zeroth-order approximation

TABLE II

Kind of pair	Number of pairs	Energy per pair	Energy of all pairs
v–v	$\frac{1}{2}z(N_v - \lambda)$	$-2X_v/z$	$-(N_v - \lambda)X_v$
u–u	$\frac{1}{2}z(N_u - \lambda)$	$-2X_u/z$	$-(N_u - \lambda)X_u$
v–u	$z\lambda$	$(-X_v - X_u + \omega)/z$	$-(X_v - X_u + \omega)\lambda$
All	$\frac{1}{2}z(N_v + N_u)$		$-N_vX_v - N_uX_u + \lambda\omega$

consists of assuming that λ is given by the random mixing value $\bar{\lambda} = N_u N_v / (N_u + N_v)$ so that the partition function becomes

$$Q(v, T) = Q^i(v, T) \frac{(N_v + N_u)!}{N_v! N_u!} \exp \left\{ \frac{N_v X_v + N_u X_u - (N_u N_v \omega / (N_u + N_v))}{kT} \right\}$$

(63)

If the $Q^i(v, T)$ can be again approximated by Eq. (54), this partition function leads to the chemical potential

$$\mu_u = \bar{E}_u^\infty - T\bar{S}_u^v + kT \ln c_u + (1 - c_u)^2 \omega$$

(64)

and the activity coefficient of the solute species is

$$\ln \gamma_u = \frac{(1 - c_u)^2 \omega}{kT} + \frac{\bar{E}_u^\infty - E_u^\circ}{kT} + \frac{\bar{S}_u^v - S_u^\circ}{k}$$

(65)

The last equation shows the relationship between the zeroth approximation and the parabolic representation of Eq. (27).

In the case of interstitial solid solutions the chemical potential of the solute species takes the form (Alex and McLellan, 1970)

$$\mu_u = \bar{E}_u^\infty - T\bar{S}_u^v + kT \ln \frac{\theta_u / \beta}{1 - (\theta_u / \beta)} + z' \varepsilon_{uu} \frac{\theta_u}{\beta}$$

(66)

where z' is the coordination number for interstitial sites and ε_{uu} is the binding energy of a u–u pair. The zeroth-order approximation has been used many times. The classic example is its application by Bragg and Williams (1934, 1935) to the superlattice problem. The approximation has also been used extensively in the discussion of ternary solid solutions where one component dissolves interstitially. It has been shown that, for such solutions, the zeroth approximation can be written (Chraska and McLellan, 1970) in the form

$$\mu_i = kT \ln \frac{\theta_i / \beta}{1 - (\theta_i / \beta)} + z' \frac{\theta_i}{\beta} \varepsilon_{ii} + \bar{E}_i^{v-i} \theta_v + \bar{E}_i^{u-i} \theta_u - T\bar{S}_i^v$$

(67)

In this equation i denotes the interstitial species and v–u is the substitutional "solvent." The quantities $\bar{E}_i^{v-i}$ and $\bar{E}_i^{u-i}$ denote the partial energies of the i atoms, in the limit of infinite dilution, in the binary v–i and u–i systems and the concentration units are

$$\theta_i = N_i / (N_u + N_v); \qquad \theta_u = N_u / (N_u + N_v); \qquad \theta_v = N_v / (N_u + N_v)$$

(68)

Equation (67) has been shown by Chraska and McLellan (1970) to be in reasonable accord with experimental data on the Ni–Cu–C system. Similar zeroth-order treatments have been developed by Kirkaldy and Purdy (1962) and Wada and Saito (1961) and applied to iron-based solutions containing carbon and a substitutional solute element.

C. The First-Order Approximation

The first-order approximation is an attempt to solve the configurational problem in a more realistic manner than the simple assumption of random solute distribution. The most familiar formulation of the first-order approximation is the quasi-chemical treatment of Guggenheim (1959). This is based on writing an initial degeneracy for the solution crystal in terms of the pairs enumerated in Table II in the form

$$g'(N_v N_u \lambda) = \frac{\{\tfrac{1}{2}z(N_v + N_u)\}! \, 2^{z\lambda}}{\{\tfrac{1}{2}z(N_v - \lambda)\}! \{\tfrac{1}{2}z(N_u - \lambda)\}! \, (z\lambda)!} \tag{69}$$

This degeneracy is inexact since the pairs in fact interact and are not independent. Secondly, a summation of $g'(N_v N_u \lambda)$ over all λ does not lead to the total possible number of distributions in the assembly. The first problem is an inherent defect of the approximation, but the second problem can be eradicated by multiplying $g'(N_v N_u \lambda)$ by a normalizing factor $h(N_v N_u)$, independent of λ, such that a summation of $g'(N_v N_u \lambda) h(N_v N_u)$ over all λ is constrained to be equal to the total number of complexions

$$\sum_\lambda g'(N_v N_u \lambda) h(N_v N_u) = (N_v + N_u)!/N_v! \, N_u! \tag{70}$$

The summation in this equation can be replaced by its greatest term. This value of λ maximizing $g'(N_v N_u \lambda)$ is the random solution value

$$\lambda^* = N_v N_u/(N_v + N_u) \tag{71}$$

This allows the value of $h(N_v N_u)$ to be found and the partition function to be written in the form

$$Q(v, T) = Q^i(v, T) \sum_\lambda \frac{(N_v + N_u)!}{N_v! \, N_u!} \cdot \frac{\{\tfrac{1}{2}z(N_v - \lambda^*)\}! \{\tfrac{1}{2}z(N_u - \lambda^*)\}! \, (z\lambda^*)! \, 2^{z\lambda}}{\{\tfrac{1}{2}z(N_v - \lambda)\}! \{\tfrac{1}{2}z(N_u - \lambda)\}! \, (z\lambda)! \, 2^{z\lambda^*}}$$
$$\times \exp\left|\frac{N_v X_v + N_u X_u - \lambda\omega}{kT}\right| \tag{72}$$

Again, in replacing the summation by its largest term, it is found that $\bar\lambda$ the value of λ maximizing $Q(v, T)$ is given by the equation,

$$(\bar\lambda)^2 = (N_v - \bar\lambda)(N_u - \bar\lambda)e^{-2\omega/zkT} \tag{73}$$

Thus finding the value of $\bar{\lambda}$ determines the configurational thermodynamic properties of the system in terms of the interchange energy ω. Various refinements of this approach have been made. Instead of simple pairs, triangular triplets or tetrahedral quadruplets have been considered (Guggenheim, 1959). In the case of substitutional solutions, however, the degree of agreement between the predictions of first-order calculations and experimental data is not impressive except in certain cases, such as the Cu–Au system. Oriani (1957) has given a critical review of the situation.

For interstitial solid solutions, the equation for $\bar{\lambda}$ (McLellan and Dunn, 1969) is

$$(\bar{\lambda})^2/(N_v\beta - N_u - \bar{\lambda})(N_u - \bar{\lambda}) = \exp\left(\varepsilon_{uu}/kT\right)$$

leading to a chemical potential

$$\mu_u = \bar{E}_u^\infty - T\bar{S}_u^v + kT \ln \frac{\theta_u/\beta}{1 - (\theta_u/\beta)}$$

$$- \frac{z'kT}{2} \ln\left(\frac{\theta_u/\beta}{1 - (\theta_u/\beta)}\right)^2 \left(\frac{1 - (\theta_u/\beta) - \phi}{(\theta_u/\beta) - \phi}\right)\exp(-\varepsilon_{uu}/kT) \qquad (74)$$

The quantity ϕ is a function of composition given by

$$\phi = \frac{1 - \{1 - 4\delta(\theta_u/\beta)(1 - (\theta_u/\beta))\}^{1/2}}{2\delta} \qquad (75)$$

where $\delta = 1 - \exp\left(-\varepsilon_{uu}/kT\right)$ and the partial energy and entropy of the solute species is given by (McLellan and Dunn, 1970)

$$\frac{\bar{S}_u}{k} = \frac{\bar{S}_u^v}{k} + \ln\left\{\left(\frac{\theta_u/\beta}{1 - (\theta_u/\beta)}\right)^{(z'-1)} \left(\frac{1 - (\theta_u/\beta) - \phi}{(\theta_u/\beta) - \phi}\right)^{z'/2}\right\}$$

$$+ \frac{z'\varepsilon_{uu}}{2kT}\left\{\frac{\exp(-\varepsilon_{uu}/kT)[(\theta_u/\beta)(1 - (\theta_u/\beta))]^2}{2\phi\delta - 1}\right\}$$

$$\times \left\{\frac{1 - 2(\theta_u/\beta)}{(\theta_u/\beta)(1 - (\theta_u/\beta)) - \phi(1 - \phi)}\right\} \qquad (76)$$

and

$$\bar{E}_u - \bar{E}_u^\infty = \frac{z'\varepsilon_{uu}}{2}\left\{1 + \left[\frac{\exp(-\varepsilon_{uu}/kT)[(\theta_u/\beta)(1 - (\theta_u/\beta))]^2}{2\phi\delta - 1}\right.\right.$$

$$\left.\left.\times \frac{1 - (2\theta_u/\beta)}{(\theta_u/\beta)(1 - (\theta_u/\beta)) - \phi(1 - \phi)}\right]\right\} \qquad (77)$$

Taking a value of 2.0 kcal/mole for ε_{uu}, the variation of $\bar{S}_u$ and $\bar{E}_u - \bar{E}_u^\infty$ with composition is shown in Figs. 6 and 7. For simplicity we have put $\beta = 1$,

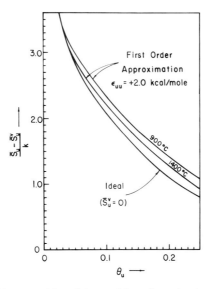

Fig. 6. Variation with composition of the partial configurational entropy in the first-order approximation for $\varepsilon_{uu} = 2.0$ kcal/mole.

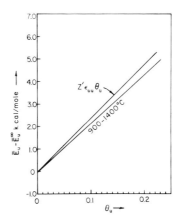

Fig. 7. Variation with composition of the partial energy in the first-order approximation for $\varepsilon_{uu} = 2.0$ kcal/mole.

corresponding to the solute atoms being located in the octahedral sites in an fcc solvent lattice.

It is clear from Figs. 6 and 7 that in the high temperature limit, the first-order model approaches the zeroth-order approximation. The upper line in Fig. 7 gives the value of $\bar{E}_u - \bar{E}_u^\infty$ calculated from the zeroth-order equation

$$\bar{E}_u - \bar{E}_u^\infty = z'\varepsilon_{uu}(\theta_u/\beta)$$

Despite the rather complicated form of Eq. (77) the values of $\bar{E}_u - \bar{E}_u^\infty$ are virtually independent of temperature in the range 900–1400°C. McLellan and Dunn (1970) have shown that the behavior of carbon in austenite can be adequately described by the first-order treatment. In this context it is interesting to note that when θ_u is small, the first-order formulation leads to the result

$$\ln \frac{\alpha_u}{\theta_u} = \theta_u\{(z' + 1) - z' \exp\left(-\varepsilon_{uu}/kT\right)\} + \frac{\bar{E}_u^\infty}{kT} - \frac{\bar{S}_u^v}{k} \tag{78}$$

which is identical in form to an expression previously obtained by Darken (1946) from a model in which each C atom is constrained to have either one or no nearest-neighbor C atoms.

In the case of ternary solid solutions, it has been shown by Stringfellow and Greene (1969) that if the three species of atom occupy the same space lattice, the number of geometrical constraints on the system is not sufficient to allow the equations of quasi-chemical equilibrium to be solved. However, when one species is dissolved on a different sublattice, the equations can be solved in a closed form. Such a solution can be represented by the diagram in Fig. 8. Again the substitutional solute atoms will be designated v and u

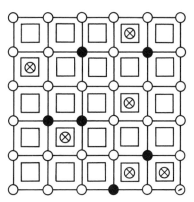

Fig. 8. Diagramatic representation of the v–u–i Ternary Solution: ○ solvent (v) atoms; ● substitutional solute (u) atoms; ⊗ interstitial atoms; □ interstitial sites.

and the interstitial atoms by i. The concentration units are those defined in Eq. (68). The total number of interstitial sites is $(N_v + N_u)\beta$. z' again denotes the coordination number for interstitial sites and z'' denotes the number of substitutional atom sites on the v–u lattice that are nearest neighbors to an interstitial site. In order to calculate the thermodynamic properties of the i atoms we need only consider the interactions between the i atoms and their v and u type neighbors and interactions between i atoms themselves. An interstitial solute will have interactions ε_{iv} and ε_{iu} with the nearest neighbor v and u atoms. When two interstitial atoms occupy adjacent sites in the lattice gas of such sites, there is an additional interaction ε_{ii}. Introducing the variables λ_1 and λ_2 enables the various pair configurations on the v–u sublattice and the interstitial site sublattice to be enumerated as in Tables III and IV. The symbol e denotes an unoccupied interstitial site.

TABLE III

Kind of pair	Number of pairs
i–u	$z''\lambda_2$
i–v	$z''(N_i - \lambda_2)$
e–u	$z''(N_u - \lambda_2)$
e–v	$z''(N_v - N_i + \lambda_2)$
All	$z''(N_v + N_u)$

Alex and McLellan (1971a) have shown that, using the geometrical constraints imposed by the crystal, the energy can be written in the form

$$E = \tfrac{1}{2}z'(N_i - \lambda_1)\varepsilon_{ii} + z''\lambda_2\,\Delta\varepsilon + z''N_i\varepsilon_{iv} \qquad (79)$$

TABLE IV

Kind of pair	Number of pairs
i–i	$\tfrac{1}{2}z'(N_i - \lambda_1)$
i–e	$\tfrac{1}{2}z'\lambda_1$
e–i	$\tfrac{1}{2}z'\lambda_1$
e–e	$\tfrac{1}{2}z'[(N_v + N_u)\beta - N_i - \lambda_1]$
All	$\tfrac{1}{2}z'(N_v + N_u)\beta$

where $\Delta\varepsilon = \varepsilon_{iu} - \varepsilon_{iv}$. The partition function is

$$Q(v, T) = Q^i(v, T) \sum_{\lambda_1,\,\lambda_2} g(N_v N_u N_i \lambda_1 \lambda_2) \exp\left(-E/kT\right) \qquad (80)$$

The procedure used to evaluate $Q(v, T)$ follows the same lines as in the case of the binary solution except that the maximum term in the summation is now determined by the two conditions

$$(\partial \ln Q(v, T)/\partial \lambda_1)_{\lambda_2} = 0 \quad \text{and} \quad (\partial \ln Q(v, T)/\partial \lambda_2)_{\lambda_1} = 0 \quad (81)$$

and the values of λ_1 and λ_2 maximizing $Q(v, T)$ are $\bar{\lambda}_1$ and $\bar{\lambda}_2$. The first condition leads to

$$(\bar{\lambda}_1)^2/((N_v + N_u)\beta - N_i - \bar{\lambda}_1)(N_i - \bar{\lambda}_1) = \exp(\varepsilon_{ii}/kT) \quad (82)$$

and the second to

$$(N_v - N_i + \bar{\lambda}_2)\bar{\lambda}_2/(N_u - \bar{\lambda}_2)(N_v + N_u - \bar{\lambda}_2) = e^{-\Delta\varepsilon/kT} \quad (83)$$

These equations can be solved and a closed expression found for $Q(v, T)$. The chemical potential of i resulting from this is

$$\mu_i = \bar{E}_i^\infty - T\bar{S}_i^v + kT \ln \frac{\theta_i/\beta}{1 - (\theta_i/\beta)}$$

$$- \frac{z'kT}{2} \ln\left(\frac{\theta_i/\beta}{1 - (\theta_i/\beta)}\right)^2 \left(\frac{1 - (\theta_i/\beta) - \phi_1}{(\theta_i/\beta) - \phi_1}\right) \exp(-\varepsilon_{ii}/kT)$$

$$- z''kT \ln \frac{\theta_i}{1 - \theta_i}\left(\frac{1 - \theta_u - \theta_i + \phi_2}{\theta_i - \phi_2}\right) \quad (84)$$

where ϕ_1 is identical to the ϕ given by Eq. (75) and replacing ε_{uu} by ε_{ii}. The function ϕ_2 is given by

$$\phi_2 = \frac{1 + (\theta_u + \theta_i)(\sigma - 1) - [\{1 + (\theta_u + \theta_i)(\delta - 1)\}^2 - 4\theta_u\theta_i\sigma(\sigma - 1)]^{1/2}}{2(\sigma - 1)}$$

$$(85)$$

where

$$\sigma = e^{-\Delta\varepsilon/kT} \quad (86)$$

Note that the infinitely dilute solution values $\bar{E}_i^\infty$ and $\bar{S}_i^v$ appearing in Eq. (84) refer to the infinitely dilute v–i binary system.

The thermodynamic functions obtained from this ternary first-order formulation have been shown to be consistent with the measured activity data for a number of ternary austenites (Alex and McLellan, 1971a). Using the value of $\varepsilon_{ii} = 2$ kcal/mole, found to be appropriate for the Fe–C binary austenite solution (McLellan and Dunn, 1970), the ternary austenite data were found to be consistent with Eq. (84) with the values of $\Delta\varepsilon$ given in Table V.

The negative values of $\Delta\varepsilon$ in Table V indicate a tendency for the C atoms to cluster round the u-type solute atoms. This is manifested by a decrease in α_i on adding component u to the solution. On the other hand, the

TABLE V

Solute (u)	$\Delta\varepsilon$ (kcal/mole)
Co	1.52
Al	−0.18
Mn	−0.93
Cr	−2.99

positive value of $\Delta\varepsilon$ in the case of Co shows that the C atoms are more strongly bonded to the Fe atoms and tend to avoid the Co atoms. As Co is added to the solution, there is an increase in α_i. It should be noted that Si greatly increases the carbon activity in ternary Fe–Si–C austenites. However, the first-order model is not capable of explaining either the older data nor the more recent data for both bcc and fcc solutions of Chraska and McLellan (1971).

It will be possible to give a much more realistic test of the models proposed for ternary solid solutions when enough data, spanning wide ranges of both solute concentration and temperature, has become available, so that the variation of $\bar{H}_u$ and $\bar{S}_u$ with solute concentration can be extracted from the data. It is always possible that theoretical expressions for μ_u, α_u, or γ_u may be in good agreement with experimental data for these quantities, whereas the partial thermodynamic properties calculated from the model may either be unreasonable or not in accord with experimental values of $\bar{E}_u$ or $\bar{S}_u$ (if such data are available).

D. Second-Order Approximations

A serious deficiency in the first-order model for solutions is the fact that the degeneracy of the maximum term in the partition function of the solution crystal is not correctly counted, leading to an error in the configurational entropy. Of course, in the high-temperature limit where $kT \gg \varepsilon_{uu}$ the error can become very small as the TS term in the free energy begins to dominate and the system approaches a random distribution.

Under some circumstances, however, the inherent counting difficulties in the quasi-chemical approach can be obviated. Kirkwood (1938) pointed out that $Q(v, T)$ can, in principal, be calculated to any desired degree of accuracy as a power series in ε_{uu}/kT. For a v–u binary substitutional solution,

$$Q(v, T) = Q^i(v, T) \exp\left(-\frac{E_0}{kT}\right) \frac{(N_v + N_u)!}{N_u! \, N_u!}$$

$$\times \left\{ 1 + \alpha M_1 + \frac{\alpha^2}{2!} M_2 + \frac{\alpha^3}{3!} M_3 + \cdots \right\} \tag{87}$$

where

$$\alpha = -\varepsilon_{uu}/kT \qquad (88)$$

and the semiinvariant moments M_i are the a priori averages over all configurations of the ith power of N_{uu}, the number of u–u nearest neighbors. Thus,

$$M_i = \frac{N_v! N_u!}{(N_u + N_v)!} \sum_c (N_{uu})^i \equiv \langle N_{uu}^i \rangle \qquad (89)$$

and the quantity E_0 is an appropriate ground state energy. The Helmholtz free energy can be found by rearranging the partition function (87) in a power series in α in the form

$$F = F^\infty - kT \ln \frac{N_v! N_u!}{(N_u + N_v)!} + kT \left\{ \alpha\lambda_1 + \frac{\alpha^2}{2!}\lambda_2 + \frac{\alpha^3}{3!}\lambda_3 + \cdots \right\} \qquad (90)$$

where the λ's are functions of the moments M_i of the form

$$\lambda_1 = M_1, \qquad\qquad \lambda_3 = M_3 - 3M_2 M_1 + 2M_1^3$$
$$\lambda_2 = M_2 - M_1^2, \qquad \lambda_4 = M_4 - 4M_3 M_1 + 3M_2^2 + 12M_1^2 M_2 - 6M_1^4$$

and so on. The moments M_i have been calculated by Bethe and Kirkwood (1939) (up to λ_4) and by Chang (1941) (λ_5) for substitutional systems and by Alex and McLellan (1971b) (up to λ_5) for interstitial systems.

This technique of calculating free energies is probably more appropriate for interstitial solutions since the solute concentration is often restricted to low values and the series in Eq. (90) converges quite rapidly as the λ_i are functions of concentration such that, for a given α, the contributions to F from successive moments decrease rapidly with decreasing solute concentration.

The use of the semiinvariant moment expansion method also allows a further disadvantage of the simpler quasi-chemical calculations to be obviated. This is because interactions between other than first-nearest neighbor solute atoms can be taken into account. In fact, the configurational energy of the solution can be written in terms of the number of first nearest neighbors X, second nearest neighbors Y, third nearest neighbors Z, etc., in the form

$$E = E_0 + \varepsilon_{uu}(X + Yt_1 + Zt_2 + \cdots) \qquad (91)$$

and the moments are now defined by

$$M_i = \langle (X + Yt_1 + Zt_2 + \cdots)^i \rangle \qquad (92)$$

where t_1, t_2, ... denote the strengths of the second, third, ... nearest neighbor pairwise interactions. The evaluation of the moments M_i and cumulants λ_i is not as difficult as it may appear. This lies in the fact that the lattice gas of solute atom sites can always be decomposed into simpler interpenetrating equivalent sublattices. Since the averages taken are a priori averages over all complexions in which all states are weighted equally, the distributions of solute atoms on two different sublattices are independent. Furthermore, if we consider the example of interstitial solute atoms located in the octahedral interstitial sites in an fcc solvent lattice, the lattice gas of such sites is itself fcc. This is shown in Fig. 9. A perusal of this figure shows

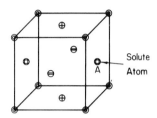

Simple cubic sublattice	First neighbor of A	Second neighbor of A	Third neighbor of A
○	4	0	8
⊕	4	0	8
◐	0	6	0
⊖	4	0	8
Total	12	6	24

Fig. 9. Decomposition of the interstitial sites in an fcc lattice into sublattices.

that the *first* nearest neighbors of any interstitial site occur on different sublattices than that on which the given site is located. The *second* nearest neighbors of a given site are all located on the same sublattice as the given site, and the *third* nearest neighbors of the given site are located on the same sublattice as the *first* nearest neighbors. This "splitting" is of great help in calculating the moments. The details of these calculations have been given in a series of papers by McLellan (1972, 1973), and it will be sufficient to state some of the results here.

The variation of the partial configurational entropy $\bar{S}_u^c$ with composition is depicted in Fig. 10. For the sake of comparison, the curve for the ideal solid solution is also included. Values of $t_1 = 0.1$ and $t_2 = 0.05$ and $T = 1000°C$ were chosen. It can be seen that, for repulsive solute mutual interactions, even when $\varepsilon_{uu} = 3$ kcal/mole, the resulting tendency toward a partial state of ordering has only a small effect on the configurational entropy. On the other hand, computing $\bar{S}_u^c$ for an attractive interaction of $\varepsilon_{uu} = -2$ kcal/mole shows a considerable reduction in the configurational

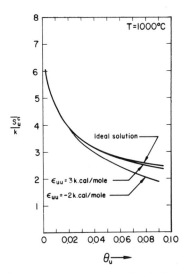

Fig. 10. Variation of $\bar{S}_u^c$ with composition calculated from the cumulant expansion technique. First, second, and third nearest neighbor solute interactions have been considered; $t_1 = 0.1$ and $t_2 = 0.05$; $T = 1000°C$.

entropy due to the tendency toward cluster formation. This result is to be expected and can be compared directly to the effect of the formation of interstitial solute atom clusters surrounding substitutional solute atoms in ternary solid solutions. It has been shown that this kind of clustering decreases the partial configurational entropy of the interstitial solute species much more than the decrease caused by the ordering tendency due to repulsive interactions between the interstitial and substitutional solute atoms (e.g., Alex and McLellan, 1971a).

The variation of $\bar{E}_u - \bar{E}_u^\infty$ with composition is shown in Fig. 11. The results are again compared with the zeroth-order model. It is seen that second and third nearest neighbor interactions have a measurable effect on the energy of solution. Similar plots for the configurational entropy showing the cumulative effect of the sets of neighbors shows that the second and third nearest neighbors have an insignificant effect on the entropy. Figure 12 shows that the energy of solution, as in the simpler models, is relatively insensitive to temperature. It should, however, be pointed out that these calculations refer to the configurational properties of the system. It is entirely possible that nonconfigurational contributions to the free energy of the solution may vary with temperature.

Low temperature expansions are much more difficult to handle on account of convergence problems. An excellent account of the problems involved has been given by Rice (1967).

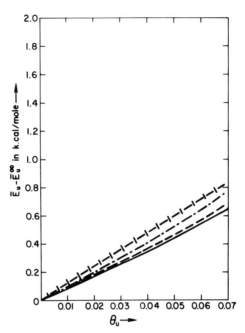

Fig. 11. Variation of $\bar{E}_u$ with composition calculated from the cumulant expansion technique (1000 C, $\varepsilon_{uu} = 1.0$ kcal/mole). First (———), second (- - - -, $t_1 = 0.1$), and third (—·—·—·, $t_1 = 0.1$, $t_2 = 0.05$) nearest neighbor solute interactions have been considered; (—|—|—|) zeroth approximation.

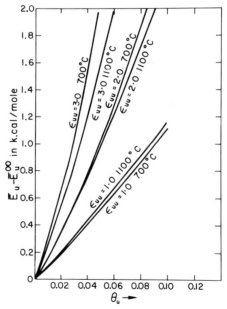

Fig. 12. Variation of $\bar{E}_u$ with temperature calculated from the cumulant expansion technique. First, second, and third nearest neighbor solute interactions have been considered. Partial energies for $t_1 = 0.1$ and $t_2 = 0.05$; units for ε_{uu} are kilocalories per mole.

E. Effects due to Volume Changes

In the preceding sections several models for solid solutions have been discussed. Many others, which are by and large modifications of those outlined, have been proposed. The modifications have not been discussed here for reasons of space. Now it is apparent that in these models it has been assumed that the molar volume of the solution is a constant. However, the bulk of the available thermodynamic data on solutions refers to measurements made at constant pressure where the molar volume changes with composition. Accordingly, as Wagner (1971) has recently pointed out, it is appropriate to consider the thermodynamic functions of the solute atoms as a function of composition at constant volume in relation to the predictions of the statistical models. It is not that the models are incorrect in calculating μ_u from the Helmholtz potential F at constant volume

$$\mu_u = (\partial F/\partial n_u)_{V, T, n_v} \tag{93}$$

but the problem lies in correlating the variation with composition of μ_u with experimental data. The chemical potential given in Eq. (93) is equal to

$$\mu_u = (\partial G/\partial n_u)_{P, T, n_v} \tag{94}$$

for any arbitrary composition but they vary with composition in a manner shown by Wagner (1971) to be related by the equation

$$(\partial \mu_u/\partial n_u)_{P, T} = (\partial \mu_u/\partial n_u)_{V, T} - (\bar{V}_u)^2/\kappa V \tag{95}$$

where κ is the isothermal compressibility,

$$\kappa = -(1/V)(\partial V/\partial P)_{T, n_u, n_v} \tag{96}$$

In the case of the infinitely dilute solution Eq. (95) is rather easy to apply because V is then known and κ refers to the pure solvent. Provided $\bar{V}_u$ is known and κ is known (or obtainable from other elastic data), Eq. (95) can be used to relate the experimental quantity $(\partial \mu_u/\partial n_u)_{P, T}$ with the calculated quantity $(\partial \mu_u/\partial n_u)_{V, T}$. By this means Wagner has shown that the activity data for C in austenite indicates that the C–C interaction energy is much stronger in the constant-volume solution than in the constant-pressure solution and that interactions between second nearest neighbor C atoms cannot be ignored.

Unfortunately, when the variation of μ_u or α_u at high solute concentrations is to be considered, we are hampered by a lack of experimental data for $\bar{V}_u$ and κ for solutions at high temperatures. The relations for $\bar{H}_u$ and $\bar{S}_u$ corresponding to Eq. (95) can also be derived.

Consider the entropy. In general, for a thermodynamic function X we can write

$$(\partial X/\partial n_u)_{P, T, n_v} = (\partial X/\partial n_u)_{V, T, n_v} + (\partial X/\partial V)_{T, n_u, n_v}(\partial V/\partial n_u)_{P, T, n_v} \quad (97)$$

Placing $X = S$ in this equation yields

$$(\partial S/\partial n_u)_{P, T, n_v} = (\partial S/\partial n_u)_{V, T, n_v} + (\partial S/\partial V)_{T, n_u, n_v} \bar{V}_u \quad (98)$$

But since

$$\left(\frac{\partial S}{\partial V}\right)_{T, n_u, n_v} = \frac{(\partial S/\partial P)_{T, n_u, n_v}}{(\partial V/\partial P)_{T, n_u, n_v}} = \frac{\alpha}{\kappa} \quad (99)$$

where α is the volume expansion coefficient, we get

$$\bar{S}_u = (\partial S/\partial n_u)_{V, T, n_v} + (\alpha/\kappa)\bar{V}_u \quad (100)$$

If we place $X = E$ in Eq. (97),

$$\bar{E}_u = (\partial E/\partial n_u)_{V, T, n_u} + (\partial E/\partial V)_{T, n_u, n_v} \bar{V}_u \quad (101)$$

Now

$$(\partial E/\partial V)_{T, n_u, n_v} = T(\partial P/\partial T)_{V, n_u, n_v} - P \quad (102)$$

but since

$$\left(\frac{\partial P}{\partial T}\right)_{V, n_u, n_v} = -\frac{(\partial V/\partial T)_{P, n_u, n_v}}{(\partial V/\partial P)_{T, n_u, n_v}} = \frac{\alpha}{\kappa} \quad (103)$$

we get

$$(\partial E/\partial V)_{T, n_u, n_v} = T(\alpha/\kappa) - P \quad (104)$$

However,

$$\bar{H}_u = (\partial E/\partial n_u)_{P, T, n_v} + P\bar{V}_u \quad (105)$$

Combining (101), (104), and (105) gives

$$\bar{H}_u = (\partial E/\partial n_u)_{V, T, n_v} + T(\alpha/\kappa)\bar{V}_u \quad (106)$$

Thus, Eqs. (100) and (106) relate the quantities $\bar{H}_u$ and $\bar{S}_u$, which can be regarded as the usual experimentally determined quantities, with the quantities $(\partial S/\partial n_u)_{V, T, n_v}$ and $(\partial E/\partial n_u)_{V, T, n_v}$, which are related more directly to constant-volume models. The most useful kind of thermodynamic data are those that yield the isothermal variation of $\bar{H}_u$ and $\bar{S}_u$ with temperature at a constant pressure. These data can be related to the partial properties calculated from constant-volume models using Eqs. (100) and (106). However, this entails a knowledge of the concentration dependence thermal and elastic data at high temperatures. Such data are scant and perhaps the better approach is to construct solution models in which volume changes are taken into consideration.

V. Models for Liquid Solutions

The general topic of models for liquids is extensive. We will attempt to cover in this short discussion only some of the statistical models or formulations that have recently been considered in respect to liquid metal solutions.

A. The Parabolic Formalism

Darken (1967) pointed out that the activity of the solute and solvent for many liquid metal solutions could be represented by Eqs. (26) and (27). The relationship between these equations and the zeroth approximation has already been pointed out. Much of the large body of activity data for solutions of liquid iron with Cu, Mn, Ni, Al, and Si seems, at the two terminal ends of the phase diagram, to be consistent with the parabolic formalism. Darken also defined an excess stability function in the form

$$\frac{d^2 G^{xs}}{dc_u^2} = \frac{1}{1 - c_u} \frac{d\bar{G}_u^{xs}}{dc_u} \tag{107}$$

This excess stability is substantially constant in the two terminal regions where the parabolic representation is found to hold. In the central region of the phase diagram, however, where departures from such simple behavior occur, the excess stability can exhibit a pronounced peak. A good example is the Mg–Bi system. At 700°C the data show reasonable compliance with Eqs. (26) and (27). In the terminal regions the excess stability is constant, but shows a well-defined peak at the composition $c_{Mg} = 0.6$. This corresponds to the composition where the compound Mg_3Bi_2 is formed in the solid alloy. Darken analyzed the data for other liquid binary systems and showed that such peaks in the excess stability functions usually occur near the composition where intermediate compounds form in the solid alloys.

Zeroth-order type approximations have also been applied to ternary liquid solutions. Gluck and Pehlke (1969) have applied Eq. (39), which assumes random mixing of the atomic species on a single lattice of sites, to the experimental activity data for dilute solutions of zinc in many different binary solvents. They found that only semiquantitative agreement existed.

B. The Quasi-chemical Approach

Although many liquid metal solutions exhibit the simple behavior reflected by the quadratic formalism, the fact remains that many more do not. There have been many attempts to apply the first-order approximation to the data for liquid metal solutions. In general, such attempts have met with only limited success. Kleppa (1957) showed that the $\bar{S}_u^{xs}$ for Zn–Sb and Cd–Sb liquids varied with solute composition in a way predicted by the quasi-chemical model. However large, but relatively constant nonconfigurational terms appear which are difficult to explain. The data for the Cu–Zn system (Downie, 1964) shows similar behavior for the entropy but the enthalpy of solution shows a composition dependence not consistent with the first-order approximation.

The application of the quasi-chemical approach in the strict sense (as defined by Guggenheim) has, in fact, found little application in the discussion of liquid metal solutions, even though the term " quasi-chemical" has been used rather loosely in this context. Several solution models that do not fit neatly into the categories described thus far have been proposed recently. Alcock and Richardson (1958, 1960) considered the solution of nonmetallic solute atoms i in a binary v–u liquid metal solvent. Their model considers the number of v–i and u–i bonds formed by the i atoms. These are related by the expression

$$N(\text{i–v})/N(\text{i–u}) = (c_v/c_u) \exp\left(-\Delta G^{xch}/kT\right) \qquad (108)$$

where ΔG^{xch} is the nonconfigurational free energy change that occurs when a u–i bond is changed to a v–i bond by the exchange of a u atom with a v atom in the nearest-neighbor coordination shell of the i atom. The application of this model to liquid metal solvents containing oxygen or sulfur atoms showed that compatibility could only be obtained with unrealistically small coordination numbers (1 or 2) for the atoms in solution. The bond model has been extended by Jacob and Alcock (1972) by assuming different bond numbers for the nonmetallic solute and the metallic solvent atoms. This modified treatment is in good accord with the experimental data for the variation of the activity coefficient of oxygen at infinite dilution with the composition of the metallic solvent. However, the model contains two adjustable parameters, n the bond number for the i species, and α, a factor related to the effect of the i atoms on the metal–metal bond strength.

Other recent models for liquid solutions have been based on the supposition of the existence of molecular species in the liquid. Belton and Tankins (1965), again in considering oxygen dissolved in liquid metals, postulated that there existed molecules of the form vi and ui in the liquid. The molecules have a random arrangement in the liquid and their relative abundance is given by an expression similar to (108). The model predicts positive excess entropies of about the right value for the Cu–Co–O system. An alternative "molecule" model has been given by Jacob and Jeffes (1971) based on the existence of species of the kind $u_2 i$ and $v_2 i$ in the liquid.

If the nonmetallic solutes such as C, N, O, B, H, and S can be regarded as dissolving interstitially in the liquid quasi lattice, then the use of Eq. (84) is immediately suggested. If the nonmetallic species is present only to a limited extent, we can ignore i–i interactions. In the limit when $\varepsilon_{ii} \to 0$, the second logarithmic term in (84) disappears and the ratio of the activity of i in the ternary solution to that in the v–i binary solution of the same i concentration becomes

$$\frac{\gamma_i^t}{\gamma_i^b} = \lim_{\theta_i \to 0} \left\{ \frac{\theta_i}{1 - \theta_i} \left(\frac{1 - \theta_u - \theta_i - \phi_2}{\theta_i - \phi_2} \right) \right\}^{-z''} \tag{109}$$

The quantity γ_i^t/γ_i^b has been designated $f_i^{(u)}$ by several authors (Fuwa and Chipman, 1959; Schürmann and Kramer, 1969) and can be obtained from (109) by the use of L'Hospital's rule. The result is

$$f_i^{(u)} = [1 - \theta_u(1 - \delta)]^{-z''} \tag{110}$$

It is easy to show that in the limit when $\theta_u \to 0$ (pure v) and $\theta_u \to 1$ (pure u) Eq. (110) gives the correct results for the pure binary v–i and u–i systems. Noting that $\Delta\varepsilon$ is really a free-energy difference, we have

$$\delta = \exp\left(-\frac{\Delta\varepsilon}{kT}\right) = \exp\left\{-\left[\frac{\Delta\bar{G}_i(u) - \Delta\bar{G}_i(v)}{z''kT}\right]\right\} \tag{111}$$

The quantities $\Delta\bar{G}_i(u)$ and $\Delta\bar{G}_i(v)$ can be regarded as experimentally known from measurements on the v–i and u–i binary systems. Values of $\Delta\varepsilon$ calculated from binary system data for several solutions containing oxygen are given in Table VI. Equation (110), using the $\Delta\varepsilon$ values of Table VI and $z'' = 6$ (fcc lattice), agrees well with the experimental data for $f_i^{(u)}$.

TABLE VI

System	$\Delta\varepsilon$ (kcal/mole)
Cu–Ag–O	2.77
Cu–Sn–O	−2.81
Ag–Pb–O	−3.79
Ag–Sn–O	−5.52

Shimoji and Niwa (1957) have presented a model for liquid metal solutions based on the cell model for liquids. Two force models were considered: one was a hard-sphere interaction with an attractive potential, and the other was a Morse potential. Only the excess thermodynamic properties were calculated. The lack of agreement with the experimental data is not surprising since the random mixing approximation was used to calculate the configurational properties.

VI. Hydrogen–Metal Solutions

The fact that hydrogen is generally a highly mobile solute element in many metals enables solubility and diffusivity measurements in metal–hydrogen systems to be made over large temperature ranges. There is, therefore, an enormous body of thermodynamic data relating to metal–hydrogen systems. In some metals the hydrogen solubility with respect to H_2 gas at atmospheric pressure is in the ppm range, whereas in other cases solubilities of $\theta \approx 1$ are obtained.

A. Data Correlations

Gallagher and Oates (1969) have noted an interesting correlation between $\Delta \bar{H}_u$ and $\Delta \bar{S}_u^{xs}$ for H–metal solid solutions. They noted that plots of $\Delta \bar{H}_u$ versus $\Delta \bar{S}_u^{xs}$ for infinitely dilute solutions were linear and fell into two groups according to whether the solvent metal was bcc or fcc (cph). There is a large amount of scatter in the correlation. Some of this may reflect the fact that the relative partial molar quantities are plotted. These quantities are normally calculated from experimental data using $H_{H_2}^\circ$ and $S_{H_2}^\circ$ for the mean temperature of the experiment. Since both $H_{H_2}^\circ$ and $S_{H_2}^\circ$ vary considerably with temperature, especially at low temperatures, the variable standard state will be reflected in the correlation. The $\Delta \bar{H}_u - \Delta \bar{S}_u^{xs}$ correlation is depicted in Fig. 13. This differs somewhat from the original figure of Gallagher and Oates in that the more recent data for Rh, Ru, Ir, and Ni (Oates and McLellan, 1973) and more recent data for Cu, Ag, and Au (McLellan, 1973) have been substituted for the older data. The newer data are indicated by squares. Using standard tables for $S_{H_2}^\circ$ the values of $\bar{S}_u^{xs}$ have been calculated for several solvent metals. The results are given in Table VII. It is interesting to note that $\bar{S}_u^{xs}$ is quite constant for the metals in Table VII with the notable exception of gold, for which $\bar{S}_u^{xs}$ is essentially zero. The data for the Au–H system (McLellan, 1973) are in substantial accord with the older measurements of Thomas (1967).

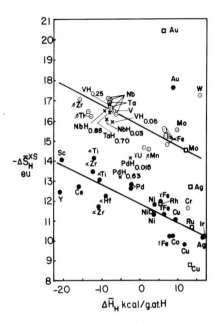

Fig. 13. Correlation between $\Delta \bar{S}_H^{xs}$ and $\Delta \bar{H}_H$ for hydrogen–metal systems.

There is also a fairly general correlation between $\Delta \bar{H}_u^\infty$ and group number for the first three long periods in the periodic table. Since $\Delta \bar{S}_u^{xs}$ is approximately the same for most metals, $\Delta \bar{H}_u^\infty$ is a reasonable index of the ability of the metal to absorb hydrogen unencumbered with complications due to hydride formation. The correlation, depicted in Fig. 14 shows the same pattern for all three long periods. There is a high maximum at group II,

TABLE VII

Solvent metal	$-\Delta \bar{S}_u^{xs}$ (e.u.)	T Range (°C)	$\dfrac{\bar{S}_u^{xs}}{k}$
Au	20.46	693–1050	−0.06
Ag	12.69	703–941	3.75
Cu	8.87	594–1027	5.63
Rh	11.99	858–1542	4.65
Ru	10.84	1002–1503	5.29
α-Ti	13.09	480–950	3.33
Co	10.27	600–1200	5.06
Ir	10.21	1393–1581	5.54

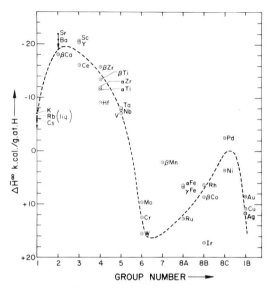

Fig. 14. Correlation between $\Delta \bar{H}_H^\infty$ and the position of the solvent metal in the periodic table.

followed by a steep drop to a minimum at group VI and then another maximum at group VIII(c), again followed by a steep decrease in solubility.

It is clear from the trends shown in Fig. 14 that it is not necessary to have an unfilled d band in order to have a high H solubility. Despite the complexities of the problem, the treatment of Ebisuzaki and O'Keefe (1967) in which the energy of solution of H in metals is calculated on the basis of a screened proton in a uniform Fermi gas has met with some success. In this model the variation in $\Delta \bar{H}_u^\infty$ from metal to metal is principally determined by the variation of the energy change occurring in the polarization of the metallic electrons by the protons. It has been pointed out by Burch that this energy varies with the density of states at the Fermi level $N(E)$. Burch (1970) has shown that $\Delta \bar{H}_u^\infty$ can be correlated with $N(E)$. However, the screened proton model also shows many features in contradiction to the thermodynamic data (Burch, 1970).

B. Location of H Atoms in Metals

The location of the dissolved H atom (or proton) in the metal lattice has been the object of much discussion and considerable research.

In the case of bcc metals the neutron diffraction studies of Wallace (1961), Rundle *et al.* (1952), and Sidhu *et al.* (1956, 1959) indicate that the tetrahedral sites are occupied by H atoms. Recent quasi-elastic neutron scattering work using niobium single crystals (Stump, 1972) also indicates that the tetrahedral sites are favored by H atoms. In addition to this evidence, recent fast ion channeling work by Carstanjen and Sizmann (1972) using niobium containing 2% deuterium shows that the D atoms occupy the tetrahedral sites.

For fcc metals the evidence is in favor of octahedral occupancy. The neutron diffraction studies of Woolan et al. (1963) on Ni–H and those of Worsham et al. (1957) on Pd–H, both show that the octahedral interstitial sites are occupied by H atoms.

It should, however, be pointed out that there is no reason to suppose that the H atoms in a given solvent crystal are completely restricted to one kind of site. Even though one kind of site is predominantly occupied, the presence of a small concentration of solute atoms on a different sublattice causes a large increase in configurational entropy (see McLellan et al., 1965).

C. Statistical Models

The simplest model that has been used in relation to H–metal solutions is the quasi-regular solution. McLellan (1969) used Eq. (61) for the chemical potential of the H atoms in the solid and Eq. (48) for those in the gas phase. Equating these two chemical potentials leads to the following equation for the solubility θ of hydrogen in the metal in equilibrium with H_2 gas at a pressure P_{H_2}:

$$\frac{\theta/\beta}{1 - (\theta/\beta)} = \frac{P_{H_2}^{1/2}\lambda}{T^{7/4}} \exp\left(-\frac{\bar{E}_H - \frac{1}{2}E_D^\circ}{kT}\right) \exp\frac{\bar{S}_H^v}{k} \tag{112}$$

where

$$\lambda = \left\{\frac{h^3}{(2\pi 2mk)^{3/2}} \frac{\sigma h^2}{8\pi^2 Ik} \frac{1}{k}\right\}^{1/2} \tag{113}$$

By and large, for the metals whose occlusive capacity is small, plots of $\ln\left[(\theta/\beta)/(1 - \theta/\beta)\right]\left[T^{7/4}/P_{H_2}^{1/2}\right]$ versus $1/T$ are linear. Another simple model for the metals of low occlusive capacity was proposed by Fowler and Smithells (1937). In this model the H atoms have translational freedom in a large fraction of the volume of the metal. For the H diffusivity, this concept leads to a prediction that is not in accord with experimental diffusivities. However, an interesting suggestion has been made by Ricca (1967) that in dilute α-Zr–H solutions some of the excess entropy is ascribable to translational motion of the H atoms in the metal.

When the concentration of H in the metal becomes large, departures from simple mixing behavior occur and more-sophisticated models must be invoked. Lacher (1937) proposed a model in which $\bar{E}_u$ for H in Pd is directly proportional to θ_u and the configurational entropy is random. Thus Lacher's model bears a close resemblance to the simple zeroth-order treatment. Schumacher (1964) also gave a treatment of H–metal solutions which corresponds to the zeroth-order approximation. Hoch (1964) has also given a zeroth-order treatment in which he includes interactions between second and third nearest neighbors. These models ascribe the deviations from regular mixing behavior to composition dependent enthalpies, whereas the configurational entropy is ideal.

Other models have been considered; these embody the blocking concept, leading to a nonideal configurational entropy. Gallagher *et al.* (1969) considered a model in which an interstitial solute atom excluded a certain number v of other sites from being occupied by other solute atoms. This leads to an expression for the partial configurational entropy given by

$$\bar{S}_u^c = -k \ln \frac{\theta_u/\beta}{1 - (\beta v/\theta_u)} \qquad (114)$$

In their model v is dependent on composition. Rees (1954) has also discussed Zr–H solutions in terms of a type of blocking model in which the occupation of a site produces a number of sites which energetically are different to unoccupied sites. The occupation of these then produces a further type of site.

A similar model has been considered by Buck and Alefeld (1971) specifically for bcc solvent metals. In contrast to fcc metals, interstitial atoms in both the octahedral and tetrahedral sites in bcc lattices produce atom displacements that are unsymmetrical with respect to the x, y, and z axes. This asymmetry has been discussed in detail by Alefeld and Buck (1969) and is incorporated into the model by supposing that the sites in the host lattice can be divided into cells containing an integral number v of sites such that only one site in a given cell can be occupied by a solute atom. The interchange of the particular occupied site within the cell leads to a new distinct microstate of the assembly. There is no interaction between cells, thus the populations of each type of site in the cells are equal. If we have a lattice containing $N_v \beta$ sites, the total number of cells will be $N_v \beta/v$. Placing N_u solute atoms in the cells, one to a cell, gives a degeneracy of

$$g = \frac{(N_v \beta/v)!}{(N_v \beta/v - N_u)! \, [(N_u/v)!]^v} \qquad (115)$$

Noting that,

$$N_u! \equiv v^{N_u} [(N_u/v)!]^v \qquad (116)$$

it can be shown that (115) leads to the expression (114).

The use of simple blocking models for fcc interstitial solutions has been criticized (see McLellan and Dunn, 1970). The model is, however, much more reasonable for bcc metals involving tetrahedral occupancy of the solute atoms since in this case a tetrahedral site has only four nearest neighbor sites.

The first-order model has also found application to the discussion of H–metal systems. Brodowsky (1965) has applied the quasi-chemical method to the Pd–H system.

There has been much discussion recently concerning the nature of the H–H interaction in H–metal solutions. The analysis of Hoch (1964) indicated, for the Ta–H system, that there is an attractive interaction between H atoms in solution and, at high concentrations, a preponderance of H–H pairs is formed. However, since the analysis is based on a zeroth-order approximation for the entropy, this conclusion is open to question.

Brodowsky (1965) proposed that the strain energy could be reduced by the formation of H–H pairs, thus yielding an effective attractive interaction between the H atoms in solution. Burch (1970) has calculated the strain energy change due to the insertion of a H atom using simple linear elastic theory.

It should, however, be pointed out that it is not simple to translate the measured datum $\bar{H}_u$ (and perhaps its variation with composition and temperature) into information regarding the H–H pairwise interaction only. Several factors complicate the issue. First, $\bar{H}_u$ is composed of many contributions from metal–metal interactions, and H–metal interactions, as well as the H–H interaction. Furthermore, these interactions can involve many nearest neighbor shells surrounding the dissolved H atom (Alex and McLellan, 1971b). Furthermore, the excess thermodynamic quantities may be changing with composition in a way that is not known, and finally, as Wagner (1971) has clearly pointed out, the comparison of $\bar{H}_u$ and $\bar{S}_u$ with partial quantities calculated from constant-volume models can lead to erroneous conclusions concerning the strength of the solute atom mutual interactions. Wagner (1971) has, in fact, concluded that the H–H interaction in V, Nb, and Ta is of a repulsive nature.

ACKNOWLEDGMENT

The author gratefully acknowledges the support provided by the U.S. Army Research Office, Durham, North Carolina.

References

Alefeld, G. (1969). *Phys. Status Solidi* (*B*) **32**, 67.
Alcock, C. B., and Richardson, F. D. (1958). *Acta Met.* **6**, 385.
Alcock, C. B., and Richardson, F. D. (1960). *Acta Met.* **8**, 882.

Alex, K., and McLellan, R. B. (1970). *J. Phys. Chem. Solids* **31**, 2751.
Alex, K., and McLellan, R. B. (1971a). *J. Phys. Chem. Solids* **32**, 449.
Alex, K., and McLellan, R. B. (1971b). *Acta Met.* **19**, 439.
Alex, K., and McLellan, R. B. (1972). *Acta Met.* **20**, 11.
Alex, K., and McLellan, R. B. (1973). *Acta Met.* **21**, 107.
Atkinson, D., and Bodsworth, C. (1970). *J. Iron Steel Inst., London* **208**, 587.
Ban-ya, S., Elliott, J. F., and Chipman, J. (1969). *Trans AIME* **245**, 1199.
Ban-ya, S., Elliott, J. F., and Chipman, J. (1970). *Met. Trans.* **1**, 1313.
Belton, G. R., and Tankins, E. S. (1965). *Trans. AIME* **233**, 1892.
Bethe, H. A., and Kirkwood, J. G. (1939). *J. Chem. Phys.* **7**, 578.
Bragg, W., and Williams, E. (1934). *Proc. Roy. Soc., Ser. A* **145**, 699.
Bragg, W., and Williams, E. (1935). *Proc. Roy. Soc., Ser. A* **151**, 540.
Brodowsky, H. (1965). *Z. Phys. Chem. (Frankfurt am Main)* **44**, 129.
Buck, H., and Alefeld, G. (1971). *Phys. Status Solidi (B)* **47**, 193.
Burch, R. (1970). *Trans. Faraday Soc.* **66**, 736.
Carstanjen, H. D., and Sizmann, R. (1972). *Proc. Hydrogen Metals Conf., Juelich, Germany,*
    Ber. Bunsenges. **76**, No. 8.
Chang, T. S. (1941). *J. Chem. Phys.* **9**, 169.
Chipman, J. (1948). *Discuss. Faraday Soc.* **4**, 23.
Chraska, P., and McLellan, R. B. (1970). *Mater. Sci. Eng.* **6**, 176.
Chraska, P., and McLellan, R. B. (1971). *Acta Met.* **19**, 1219.
Darken, L. S. (1946). *J. Amer. Chem. Soc.* **68**, 1163.
Darken, L. S. (1950). *J. Amer. Chem. Soc.* **72**, 2909.
Darken, L. S. (1967). *Trans. AIME* **239**, 80.
Darken, L. S., and Gurry, R. W. (1953). " Physical Chemistry of Metals." McGraw-Hill, New
    York.
Downie, D. B. (1964). *Acta Met.* **12**, 875.
Dunn, W. W., and McLellan, R. B. (1971). *Met. Trans.* **2**, 1079.
Dunn, W. W., McLellan, R. B., and Oates, W. A. (1968). *Trans. AIME* **242**, 2129.
Ebisuzaki, Y., and O'Keefe, M. O. (1967). *Progr. Solid State Chem.* **4**, 187.
Fowler, R. H., and Guggenheim, E. A. (1949). " Statistical Thermodynamics." Cambridge
    Univ. Press, London and New York.
Fowler, R. H., and Smithells, C. J. (1937). *Proc. Roy. Soc., Ser. A* **160**, 37.
Freedman, J. F., and Nowick, A. S. (1958). *Acta Met.* **6**, 176.
Fuwa, T., and Chipman, J. (1959). *Trans. AIME* **215**, 708.
Gallagher, P. T., and Oates, W. A. (1969). *Trans. AIME* **245**, 179.
Gallagher, P. T., Lambert, J. A., and Oates, W. A. (1969). *Trans. AIME* **245**, 47.
Gluck, J. V., and Pehlke, R. D. (1969). *Trans. AIME* **245**, 711.
Guggenheim, E. A. (1959). " Mixtures." Oxford Univ. Press, London and New York.
Hoch, M. (1964). *Trans. AIME* **230**, 138.
Jacob, K. T., and Alcock, C. B. (1972). *Acta Met.* **20**, 221.
Jacob, K. T., and Jeffes, J. H. E. (1971). *Inst. Mining Metal., Trans., Sect. C* **80**, 32.
Johnson, G. W., and Shuttleworth, R. (1959). *Phil. Mag.* **4**, 957.
Kirkaldy, J. S., and Purdy, G. R. (1962). *Can. J. Phys.* **40**, 202.
Kirkwood, J. G. (1938). *J. Chem. Phys.* **6**, 70.
Kleppa, O. J. (1957). *In* " Liquid Metals and Solidification," p. 307. Amer. Soc. Metals, Metals
    Park, Ohio.
Kubaschewski, O., Evans, E. L. L., and Alcock, C. B. (1967). " Metallurgical Thermochemistry."
    Pergamon, Oxford.
Lacher, J. R. (1937). *Proc. Roy. Soc., Ser. A* **161**, 523.
Lupis, C. H. P., and Elliott, J. F. (1965). *Trans. AIME* **233**, 829.

Lupis, C. H. P. and Elliott, J. F. (1966a). *Acta Met.* **14**, 529.
Lupis, C. H. P. and Elliott, J. F. (1966b). *Acta Met.* **14**, 1019.
McLellan, R. B. (1964). *Trans. AIME* **230**, 1468.
McLellan, R. B. (1965). *Trans AIME* **233**, 1664.
McLellan, R. B. (1969). *Scr. Met.* **3**, 389.
McLellan, R. B. (1969). *In* "Phase Stability in Metals and Alloys" (P. S. Rudman, J. Stringer, and R. I. Jaffee, eds), p. 393. McGraw-Hill, New York.
McLellan, R. B. (1972). *Mater. Sci. Eng.* **9**, 121.
McLellan, R. B. (1973). *J. Phys. Chem. Solids* **34**, 1137.
McLellan, R. B., and Chraska, P. (1971). *Mater. Sci. Eng.* **7**, 305.
McLellan, R. B., and Dunn, W. W. (1969). *J. Phys. Chem. Solids* **30**, 2631.
McLellan, R. B., and Dunn, W. W. (1970). *Scr. Met.* **4**, 321.
McLellan, R. B., Rudee, M. L., and Ishibachi, T. (1965). *Trans. AIME* **233**, 1938.
Moya, F., Moya-Gontier, G. E., Cabane-Brouty, F., and Oudar, J. (1971). *Acta Met.* **19**, 1189.
Oates, W. A., and McLellan, R. B. (1968). *Trans. AIME* **242**, 1477.
Oates, W. A., and McLellan, R. B. (1973). *Acta Met.* **21**, 181.
Parris, D. C., and McLellan, R. B. (1972). *Mater. Sci. Eng.* **9**, 181.
Pastorek, R. L., and Rapp, R. A. (1969). *Trans. AIME* **245**, 1711.
Rees, A. L. G. (1954). *Trans. Faraday. Soc.* **50**, 335.
Ricca, F. (1967). *J. Phys. Chem.* **71**, 3632.
Rice, O. K. (1967). "Statistical Mechanics, Thermodynamics, and Kinetics." Freeman, San Francisco, California.
Rundle, R. E., Shull, C. G., and Woolan, E. O. (1952). *Acta Crystallogr.* **5**, 22.
Scatchard, G. (1949). *Chem. Rev.* **44**, 7.
Scheil, E. (1943). *Z. Elektrochem.* **49**, 242.
Schürmann, E., and Kramer, D. (1969). *Giessereiforschung* **21**, 29.
Schumacher, D. P. (1964). *J. Chem. Phys.* **40**, 153.
Shimoji, M., and Niwa, K. (1957). *Acta Met.* **5**, 496.
Sidhu, S. S., Heaton, L., and Zauberis, D. D. (1956). *Acta Crystallogr.* **9**, 607.
Sidhu, S. S., Heaton, L., and Mueller, M. H. (1959). *J. Appl. Phys.* **30**, 1323.
Siller, R. H., Oates, W. A., and McLellan, R. B. (1968). *J. Less-Common Metals* **16**, 71.
Stringfellow, G. B., and Greene, P. E. (1969). *J. Phys. Chem. Solids* **30**, 1779.
Stump, N. (1972). *Proc. Hydrogen Metals Conf., Juelich, Germany, Ber. Bunsenges.* **76**, No. 8, pp. 782.
Thomas, C. L. (1967). *Trans. AIME* **239**, 485.
Turkdogan, E. T., and Darken, L. S. (1968). *Trans. AIME* **242**, 1997.
Turkdogan, E. T., Darken, L. S., and Fruehan, R. J. (1969). *Trans. AIME* **245**, 1003.
Wada, H., and Saito, T. (1961). *Trans. Jap. Inst. Metals* **2**, 15.
Wagner, C. (1954). *Acta Met.* **2**, 242.
Wagner, C. (1962). "Thermodynamics of Alloys." Addison-Wesley, Reading, Massachusetts.
Wagner, C. (1971). *Acta Met.* **19**, 843.
Wallace, W. E. (1961). *J. Chem. Phys.* **35**, 2156.
Woolan, E. O., Cable, J. W., and Koehler, W. C. (1963). *J. Phys. Chem. Solids* **24**, 1141.
Worsham, J. E., Wilkinson, M. K., and Shull, C. G. (1957). *J. Phys. Chem. Solids* **3**, 303.
Zupp, R. R., and Stevenson, D. A. (1966). *Trans. AIME* **242**, 862.

Radiation Studies of Materials Using Color Centers

W. A. SIBLEY

Department of Physics
Oklahoma State University
Stillwater, Oklahoma

and

DEREK POOLEY

Materials Physics Division
Atomic Energy Research Establishment
Harwell, Berkshire, United Kingdom

I. Introduction

Although there is now a strong emphasis on applied or " useful " physics, it remains widely acknowledged that long-term technological progress rests heavily on effective fundamental research programs. Three kinds of justification are often given for fundamental research in physics, and the study of radiation-induced color centers, that is defects which cause optical absorption in solids, can to some extent be justified on all three counts.

The first justification is creative, that is that the work is worth doing simply because of the insight it brings into the structure and operation of the physical world. This argument, although clearly a powerful one in cases such as high energy physics or cosmology, can scarcely justify detailed work in cataloguing large numbers of color centers in a wide range of materials. Nevertheless, color center spectroscopy does give considerable insight into the interactions between electrons and other electrons, nuclei, or photons. The second justification is that technology which already exists will need to call on the knowledge already available or expected to be generated in the fundamental program. In this vein perhaps even more important than the knowledge generated is the expertise of the people who work in the fundamental research program. Thus, it was likely from the outset of nuclear power generation that reactor materials would suffer seriously from radiation damage, and it was argued that research in radiation damage in metals and insulators should be initiated. This decision has been abundantly vindicated. Reactor materials have time and again failed because of radiation damage, and the understanding of radiation damage generated in basic research programs has been applied very effectively. The third justification is that wide-ranging research carried out by " good " scientists will often have impact on technological areas in unimagined ways, and may even generate completely new areas of technology. Color centers are already widely used in dosimetry and have a potentially wide application in information storage and processing devices. In this sense, they are good examples of how applications in diverse areas of technology are possible.

Bearing the above justifications in mind, we would argue that some emphasis must continue to be placed on research on " model " materials, which enable scientists to compare theory and experiment and deepen their understanding. Technologically important materials should never be ignored, and it is important that fundamental work should be done by people who remain alert for potential uses of their supposedly basic research.

The main aim of this chapter is to give an introduction to color center physics and show how it can be used to study radiation damage. The goal in

Section IV is to give examples of how color centers are currently being exploited technologically. Space will not allow a detailed treatment of all the intricacies, either of color center or radiation damage physics. Instead, we will attempt to make a presentation that expounds the underlying theme and refers to more-detailed reviews or to original papers where considered appropriate.

High energy radiation creates point defects in all solid materials. In metals these defects are rarely detectable except through their effects on electrical and thermal conductivity and on the mechanical properties of the material. Since these are macroscopic measurements, none of them give a clear indication of the structure of the defects or yield sufficient information to define the details of radiation damage events. In insulators, on the other hand, the defects are called color centers because they usually have a number of electronic states between which optical transitions are possible. Spectroscopic studies of color centers, using a variety of optical and magnetic resonance techniques, provide a rich source of information about defect structure and damage processes which is incomparably superior to that obtainable in metals. This is our theme.

It should be pointed out initially that it has been the experience of many of us who have worked in this field that the use of several experimental tools on one problem is not only desirable but almost a necessity. For example, in alkali halides irradiated with X rays or gamma rays at room temperature, only 5 ppm of Pb impurity is sufficient to change totally the radiation damage rate from that characteristic of a crystal that contains only a few parts per billion Pb. This means impurity analyses as well as radiation damage rate measurements are necessary. The interrelationships between dislocations, vacancies, and impurities which exist for either charge trapping or ionic defect trapping are also quite complex. Thus, it is important that experimental techniques be utilized that keep track of as many of these entities as possible. Even different measurements of the same parameter are often useful. This aspect will be emphasized in this chapter. In the case of the F center in alkali halides or the F^+ center in alkaline earth oxides, negative-ion vacancies with one trapped electron per vacancy, both optical absorption and electron paramagnetic resonance can be used to monitor the defect concentration. However, a combination of these two tools allows a determination of an oscillator strength for the optical transition, which can then be used to determine the "concentration of centers present." The main point to be kept in mind is that in order to characterize materials properly, each crystal must be studied in as much detail as possible with a variety of experimental techniques. In the future this will require much more cooperation between laboratories and scientists than has been true in the past, primarily because all research is becoming increasingly expensive.

II. The Physics of Color Centers

A. Perfect Solids

Perfect crystalline solids consist of regular, three-dimensional arrays of atoms. The chemical bonds that hold the atoms together are usually considered as falling into three broad categories: ionic, covalent, and metallic. In our study of color centers in insulating inorganic solids, we will be mostly concerned with ionic bonding. In terms of their electronic bond structure, solids are classified as insulators, semiconductors, or metals, and ionic solids are invariably insulators. It is useful at the outset to consider briefly some of the implications of bonding and electronic structure on color center physics and radiation damage.

1. BONDING

The classical examples of solids with predominantly ionic bonding are the alkali halides, such as NaCl. In a sodium atom, one electron is in a 3s orbital outside a closed electronic shell structure. The closed shell is spherically symmetric and effectively shields the outer electron from most of the positive charge of the nucleus; in fact, the outer electron "sees" only one positive charge. As a result, it is not very costly in energy terms to remove an electron from a sodium atom; its *ionization potential* is only 5.1 eV. In chlorine atoms, on the other hand, the outer electronic shell (3p) needs one extra electron to fill it. Because the other outer electrons are not able to screen the nuclear charge from the added electron, placing an electron into this "hole" in the outer shell is an energetically favorable process, in spite of the fact that the chlorine atom is neutral. In fact, when an electron is added to chlorine, there is a net energy gain; its electron affinity is 4.3 eV. As a result, the energy required to remove an electron from a sodium atom and place it on a chlorine atom (0.8 eV) is very much less than the energy gained when the now oppositely charged ions are brought together into a regular array in a solid (7.9 eV). Thus, in ionic solids it is energetically favorable for the constituent atoms to exist as positively and negatively charged ions.

In ionic solids where the ions have charges of $\pm ze$ and the nearest neighbor separation is r, the electrostatic binding energy is Az^2e^2/r, where A is called the Madelung constant. This constant depends on the structure of the crystal but is usually in the range 1.6–1.8. In alkali halides, then, where $z = 1$, the binding energy is in the region of 6–10 eV. In alkaline earth oxides such as MgO, where $z \sim 2$, the electrostatic binding energy is in the region of 30–40 eV. It is, of course, this high binding energy that is responsible for the hardness and refractory nature of MgO compared with NaCl.

The binding energy also has a profound effect on the radiation damage process, which is very different in the two materials; NaCl is very sensitive to radiation and colors easily, quite unlike MgO. In an oversimplified sense, this is obvious. The fundamental electronic excitations produced by ionizing radiation in alkali halides have energies of 5–10 eV, which is usually enough energy to create defects since the atomic binding energies are in the same energy range. In alkaline earth oxides, on the other hand, where the binding energies are 30–40 eV, the electronic excitations are still only 5–10 eV and are clearly unable to cause atomic displacement.

2. Electronic Structure

Ionic solids are good insulators because there is a 5–10 eV energy gap between the highest occupied electron band (the valence band), which is completely filled and cannot therefore conduct electricity, and the lowest unoccupied band (the conduction band), which is completely empty. We have already pointed out that, from the point of view of the energetics of radiation effects in insulators, it is important that the electron–hole pairs created when an electron is excited from the valence band to the conduction band have this fairly high energy. It is also important that the electron–hole pairs have a fairly long lifetime. A long lifetime is characteristic for the large band gap materials because the excited electron must lose a great deal of energy at once in recombining with the hole, and this is very difficult since there are no allowed levels in the band gap. In insulators, electron–hole recombination usually takes place in a time of 1 nsec to 1 μsec. This is much longer than the time required for electron or hole trapping or for atomic vibrations (10^{12} sec^{-1} or about 1 psec). Metals behave very differently. All electronic excitations in metals can decay by sharing energy solely with other electrons, with the result that they decay very quickly (< 1 psec).

It is also evident that the physics of *color* centers in insulators only exists because the large band gap in insulators makes them transparent to a wide range of visible photons. These photons do not have enough energy to excite electrons across the band gap but can excite electrons at defect sites. Metals, on the other hand, interact strongly with light at all wavelengths, from the infrared through to the ultraviolet, and as a result no color center spectroscopy is possible.

B. Defects and Their Classification

Real solids always contain both *point* defects and *extended* defects. We will be mostly concerned with the former. Conceptually, the simplest of all point defects are isolated *interstitials* or *vacancies* in elemental solids such as metals. The interstitial is an atom that occupies a position where no atom would exist in the perfect crystal, and the vacancy is where an atom is missing from the perfect lattice structure. It is worth pointing out at this stage that color center physicists usually distinguish between a missing *atom* and a missing *electron* by using the words *vacancy* and *hole*, respectively. Point defects can be much more complicated than isolated vacancies and interstitials; the phrase point defect is used for any complex of simple defects which still has a fairly simple symmetry. Figure 1a shows some point defects in a simple two-dimensional square lattice.

Extended defects can be thought of as aggregates of point defects, which are so large that simple symmetry properties have been lost. By far the most important are dislocations that consist of the boundaries of missing or extra planes of atoms (see Fig. 1b). Dislocations are crucially important in determining the mechanical properties of most solids; for example, they explain

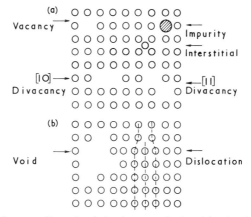

Fig. 1. Defects in a two-dimensional simple square lattice: (a) point defects; (b) extended defects.

why the critical shear stress in most materials is many orders of magnitude smaller than would be expected for perfect crystals. In *dislocation loops* the extent of the extra or missing plane of atoms is fairly small, so that the core of the dislocation (where the extra plane ends) forms a finite loop. These loops are the commonest result of point defect aggregation after radiation damage. Voids are an example of a three-dimensional defect aggregate, but we will not be concerned with them.

Defects can also be classified as intrinsic or extrinsic. *Intrinsic* defects involve only the atoms of the perfect solid and are made up by rearranging them. Simple vacancies and interstitials are examples of intrinsic defects, as are dislocation loops. If impurity atoms are involved in the defect, then it is called *extrinsic*. Very often intrinsic vacancies or interstitials, which can usually move fairly freely at high temperatures, prefer to associate with impurity atoms, so that impurity–vacancy or impurity–interstitial complexes are formed. These complexes, and of course the isolated impurity atoms themselves, are examples of simple extrinsic defects.

Defects in alkali halides, alkaline earth halides, and in simple oxides will be our main concern. It is therefore necessary to present the rules that have been evolved for naming defects in these materials. Most of the letters used to designate defects were given initially without thought of the structure, usually for the very good reason that the structure was not known at the time. Recently Sonder and Sibley (1972) have introduced some order into the resulting chaos, while at the same time retaining most of the features of the existing notation. The most important points of this notation are as follows:

(a) The basic atomic character of the defect is represented by a letter: F is used for negative ion vacancies, V for positive ion vacancies, I for interstitial species, and H for the special form of interstitial known in alkali halides where two atoms or ions share one lattice site.

(b) The electronic character of the defect is given by a superscript charge, which is the effective charge of the defect in the lattice. Thus, a negative-ion vacancy in alkali halides has an effective positive charge since a negative ion has been removed from an otherwise neutral lattice, and it is designated F^+. If an electron is trapped in the same vacancy, the defect becomes an F center.

(c) Aggregates of point defects are written as if they were molecules; for example, two neighboring F centers, which form what is traditionally called an M center, is designated F_2 in this notation.

(d) Impurity atoms are included in the notation as a subscript. An F center sitting next to an impurity alkali ion would be called an F_A center, and a positive ion vacancy next to an OH^- ion and having trapped a hole in an alkaline earth oxide would be a V_{OH} center.

(e) If the defect contains no extra or missing atoms, square brackets are used to describe the effective molecular species involved. The trapped hole in alkali halides, previously called the V_k center, is really a halogen molecule ion (Cl_2^- in NaCl) which occupies two halide ion sites. In the new notation, this center would be $[Cl_2^-]^+$, with the positive sign denoting the positive net charge of the defect.

C. Optical Properties of Color Centers

The most obvious effect of defects in insulators is on their optical properties, and because of this, optical techniques are almost invariably the ones through which an initial study of the centers is made. The optical effect of F

centers on alkali halides is spectacular; the pure single crystals are absolutely clear and colorless, but those containing F centers are often brightly
colored. In this section, we will try to show how the different features of the
optical spectra of color centers help to unravel details, both of their structure
and of the radiation damage process by which they are formed.

For the F center in alkali halides, the correct structure was first proposed
by de Boer (1937), namely a single anion vacancy with a single trapped
electron. F centers can be produced in most alkali halides by additive
coloration in excess alkali metal vapor at sufficiently high temperatures or
by irradiation with X rays or high energy particles. They have been studied
very intensively and a number of review articles that deal with their properties are now available (Schulman and Compton, 1962; Markham, 1966;
Fowler, 1968; Klick, 1972). However, although the alkali halide F center is
the simplest center of all in concept, and very widely studied, its properties
are not always simple, and it does not always lend itself to the analytical
power of a skillful use of optical spectroscopy. For this reason, we will use
not only F centers but also F_2 centers (M centers) and more complex F
aggregate and hole centers for illustration.

The M band produced in X-irradiated crystals was noticed as early as
1928 by Ottmer (1928). It can also be generated by optical bleaching of
crystals containing F centers with F band light. The letter M was given in
recognition of the work on the center by Molner (1941), but it was the work
of van Doorn (1962) that led to the assignment of the F_2 structure, that is,
two F centers on neighboring anion sites. A detailed review of work on the
M center has been given by Compton and Rabin (1964). Larger F center
aggregates are less well characterized, although the R center is known to be
an equilateral F_3 configuration, and the N center may be an F_4
configuration. The wide variety of hole centers that can be produced in
alkali halides and alkaline earth oxides by irradiation are reviewed by
Kabler (1972) and Hughes and Henderson (1972), respectively.

1. THE INTENSITY OF OPTICAL ABSORPTION BANDS

When light with photon energy E and flux $F_0(E)$ is incident on a single
crystal sample the transmitted flux F is related to F_0 by the equation

$$F/F_0 \simeq [1 - R]^2 \exp(-\alpha t) \tag{1}$$

where t is the thickness of the sample. $R(E)$ is the reflection coefficient of the
surface for normal incidence and is given by

$$R = (\eta - 1)^2/(\eta + 1)^2 \tag{2}$$

where η is the refractive index of the crystal. This is changed very little by
incorporation of color centers. On the other hand, the absorption coefficient

α is changed very significantly by color centers because it is very nearly zero for unirradiated crystals in the near uv, visible, and near ir regions. In optical measurements, it is common to use optical density (OD), which is defined as

$$OD = \log_{10}\left[F_0/F\right] = 0.4343\alpha t - 2\log_{10}\left[1 - R\right] \tag{3}$$

Since R is not much changed by color centers, it follows that the optical density changes which are measured can be directly related to changes in α.

The intensity of an optical absorption band is the area under the curve of the absorption coefficient versus photon energy, that is

$$\int_{E_{min}}^{E_{max}} \alpha(E)\, dE \tag{4}$$

Although the width of a particular band will usually change considerably with temperature (Fig. 2), the area under the band will not, unless the

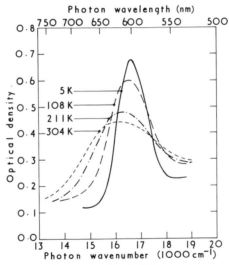

Fig. 2. The F band in an electron irradiated KBr crystal, measured at four different temperatures.

number of centers causing the band changes or unless the oscillator strength for the transition changes, which is not often the case for color centers. In fact, the area under the absorption band is usually directly proportional to the number of defects in the sample, and it can therefore be used as a measure of defect concentration.

At a given temperature, the shape of a band does not depend on the number of centers contributing to it, and even at different temperatures the maximum absorption coefficient in the band α_{max}, and its width W change together in such a way that $\alpha_{max} W$ depends only on the number of centers and not on temperature. Hence, α_{max} or $\alpha_{max} W$ can be used to give center concentration. The shape of the color center band may be lorentzian or gaussian. If it is gaussian, as most color center bands are, one can write the following relationship between number of centers and $\alpha_{max} W$:

$$Nf = 0.87 \times 10^{21} \frac{\eta}{(\eta^2 + 2)^2} \alpha_{max} W \quad m^{-3} \qquad (5)$$

This is usually called Smakula's formula. N is the concentration of centers per cubic meter, f is the oscillator strength of the absorption band, η is the refractive index at the peak of the band, α_{max} is the maximum absorption coefficient in inverse meters, and W is the band width at half-height in electron volts. For allowed electric dipole transitions, f is always near unity; it is about 0.7 for F bands in alkali halides. For alkali halides $\eta \sim 1.5$, and $W \sim 0.5$ eV for F centers so that

$$N_F \sim 10^{16} \alpha_{max} \quad cm^{-3} \qquad (6)$$

The power of this method of measuring defect concentrations can hardly be overstressed. Figure 3 shows the optical absorption spectrum of KCl measured at ~ 100 K after additive coloration at 800 K and bleaching in the F band at room temperature. In the figure, the F, R_1, R_2, M, and N bands are easily seen, and using the peak heights of the bands we can estimate the concentrations of each. In a metal, on the other hand, we would be forced to use electrical resistivity to measure defect concentration comparably simply,

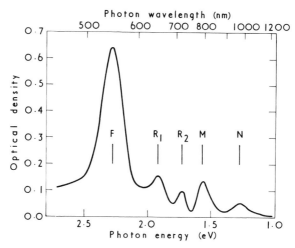

Fig. 3. F and F-aggregate bands in KCl, measured at ~ 100 K after additive coloration at 500°C and optical bleaching at room temperature. After Petroff (1950).

and it would not then be possible to separate the contributions to resistivity from different kinds of defects.

One of the more spectacular applications of optical absorption in defining the radiation damage process in alkali halides was provided by the work of Ueta *et al.* (1969). They were able to follow the growth of the F band during the period of a few microseconds following a very intense radiation pulse (Fig.4) and show that the mechanism of radiation damage must involve the creation of the F center in its ground electronic state. This information was crucial in the formulation of a mechanism for defect creation (Pooley and Hatcher, 1970; Smoluchowski *et al.*, 1971), and could only be obtained because optical absorption provided an accurate and very fast measure of F center concentration.

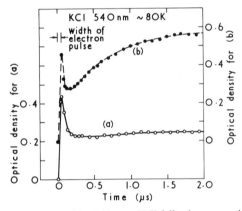

Fig. 4. The growth of the F band in KCl at ∼ 80 K following a very short intense electron irradiation pulse; (a) a freshly cleaved crystal; (b) a previously irradiated crystal. After Ueta *et al.* (1969).

Measurement of defect concentration can also be important in structure determination. For example, the structure of the M center was partly determined by showing (van Doorn, 1962) that when F and M centers are in equilibrium the concentration of the M centers varies as the square of the F center concentration (Fig. 5). This observation helped confirm that M centers consisted of two F centers.

2. The Position of Optical Absorption Bands

The F center in alkali halides is an electron trapped in an anion vacancy; it is therefore not surprising that it has some of the properties of a particle in a cubic box. For example, we might expect the transition energy between the ground state and the first excited state to vary with the inverse square of the

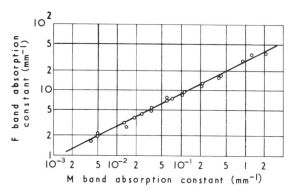

Fig. 5. The relationship between F and M center concentrations in additively colored KCl. Equilibrium was established at 697°C and optical absorption measurements made at 77 K. After van Doorn (1962).

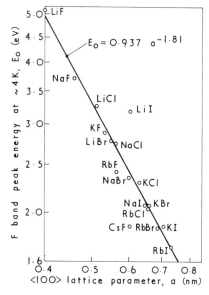

Fig. 6. The Mollwo–Ivey relationship between F band peak energies at ~ 4 K and lattice parameter for the alkali halides with the NaCl structure. After Dawson and Pooley (1969).

lattice parameter and to have threefold degeneracy. Since there are 17 alkali halides with the NaCl structure, it is fairly easy to determine experimentally how the position of the F band does vary with lattice parameter. Figure 6 shows that the prediction of the particle in the box model is not too bad. This kind of relation between energy and lattice parameter was first noticed by Mollwo (1931) and refined by Ivey (1947). Its use in predicting where an absorption band should be has proved a valuable tool in helping to identify centers in less-well-studied crystals, such as the F center in LiBr (Dawson and Pooley, 1969).

The threefold degeneracy of the first excited state of a particle in a cubic box can be lifted by an external perturbation, and in the same way, the degeneracy of the excited state of the F center can be lifted. Possibly the most spectacular perturbation is provided by a neighboring impurity ion, as in the F_A center in alkali halides (Lüty, 1968). In KCl, the F band is at 2.31 eV at ~ 5 K, whereas the F_A center has bands at 2.35 and 2.12 eV, with an intensity ratio of 2 : 1. This is just what is expected when a triply degenerate p-like excited state is highly disturbed by an axial perturbation. This observation helps confirm the structure of the F_A center. Similarly, the M and M_F bands (the M_F band absorbs light in the F band region) associated with the M center are just as expected for the F_2 structure and help to confirm the F_2 model.

These examples illustrate how the position of an optical absorption band on the photon energy scale, together with the number, relative intensity, and relative position of various components, can often be very useful in deciding what kind of defects give rise to particular bands.

3. DICHROISM

Complex centers rarely have the high symmetry of the F center, and as a result, their optical properties often depend critically on the alignment of the polarization vector of the measuring light with respect to the alignment of the center. Suppose, for example, that polarized light with a wavelength equal to that of the peak of the M band is passed through a crystal containing M centers all aligned in one direction. If the polarization vector is perpendicular to the axis of the M centers, no absorption occurs, but for polarization parallel to the centers the absorption is strong. In general, if θ is the angle between the polarization vector and the axis of the centers, then the absorption coefficient is $\alpha_{max} \cos^2 \theta$. A crystal like this, containing centers that are preferentially aligned, is termed dichroic. Obviously, dichroic effects cannot be seen with unpolarized light, nor do they exist in cubic crystals where all possible orientations of centers are equally probable, as is the case for M centers in a freshly irradiated crystal.

M center dichroism, however, is easy to create by bleaching with polarized light in the M_F band. M centers *do not* absorb M_F band light if the polarization vector of the light is parallel to the axis of the center, but if the vector has some component perpendicular to the center, absorption will occur. The centers that do absorb M_F photons have a finite probability of reorienting into another of the six possible orientations shown in Fig. 7,

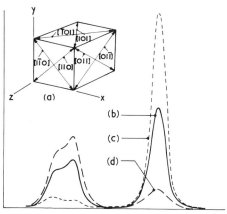

Fig. 7. M center dichroism in alkali halides: (a) the six possible ⟨110⟩ orientations of M centers; (b) the optical absorption spectrum of a freshly irradiated crystal measured with [110] polarized light incident on the [001] direction (the z axis); (c) as for (b) but after bleaching with M_F light polarized on [110]; (d) as for (c) but measured with [110] polarized light.

before relaxing into the ground state. It follows that if a crystal is irradiated with M_F light incident on the z axis of Fig. 7 and polarized on [110], then all the centers except those on the [110] axis are excited. Since all excited centers have a finite probability of relaxing into the [110] orientation, from which no return is possible since these centers are not excited, all the M centers eventually adopt this orientation. As Fig. 7c and d shows, the crystal is then highly dichroic. It is evident from this that dichroic optical absorption can help considerably in determining the symmetry of a center (e.g., Ueta, 1952).

4. Polarized Luminescence

It is not always necessary to reorient centers with low symmetry in order to show that they have directional optical properties. If suitable lumine-scence transitions occur, then directional behavior can be seen even in crystals containing equal numbers of centers of all possible orientations. Thus, in many crystals the M center luminesces at room temperature, and the electric dipole moment of the luminescence is parallel to the axis of the center as is that of the M band absorption. If we take a crystal containing M centers randomly oriented and excite it with M band light incident on a [001] direction and polarized along either [100] or [110], the intensity of the luminescence depends on the direction along which the luminescence is detected (Table I). Even more discrimination can be achieved if the luminescence is detected through a polarizer.

TABLE I

RELATIVE INTENSITIES OF LUMINESCENCE EXCITED BY POLARIZED
LIGHT INCIDENT ON AN ALKALI HALIDE CRYSTAL
ALONG THE [001] DIRECTION[a]

Exit direction	Incident light polarization	$\langle 110 \rangle$ centers (M centers)		$\langle 100 \rangle$ centers (F_A centers)	
		[100]	[110]	[100]	[110]
[100]		4	5	0	2
[010]		6	5	4	2
[001]		6	6	4	4

[a] It is assumed that the centers have excitation and luminescence dipoles parallel to their axis of symmetry and are randomly oriented.

The intensities given in Table I have been very simply calculated, making the assumption that the incident light beam is not significantly reduced in intensity because of absorption in the M centers. This means that the rate of excitation of centers is just the triple product of the intensity of the incident light, the number of centers in the particular orientation, and $\cos^2 \theta$, where θ is the angle between the polarization vector of the incident beam and the orientation direction in question. Similarly, the probability that an excited center will emit luminescence along another direction at ϕ to the direction of the center is $\sin^2 \phi$. It hardly needs emphasizing that polarized luminescence can, like dichroism, allow us to distinguish between centers of different symmetry. This is illustrated by Table I, which also shows the results of the experiments discussed above, but with F_A light and F_A centers as well as with M light and M centers.

A significant example of the use of a combination of dichroism and polarized luminescence is provided by the work of Kabler (1964) and Murray and Keller (1965) on the exciton state in alkali halides. Exciton is the name given to excited electronic states of all solids which are essentially bound electron–hole pairs, but in alkali halides excitons were shown by Kabler and Murray and Keller to be very different from the spherically symmetric hydrogen atomlike structures found in semiconductors. They were able to show that the exciton had the same $\langle 110 \rangle$ symmetry axis as the trapped hole or $[X_2^-]^+$ center. First, they made $[X_2^-]^+$ centers at $\sim 5°$K by X irradiating crystals containing impurity atoms, which were able to trap electrons and prevent their recombining with the $[X_2^-]^+$ centers. Then, the $[X_2^-]^+$ centers were oriented by exciting with polarized light in the $[X_2^-]^+$ band. It had been shown earlier by Castner and Kanzig (1957) and by Delbecq et al. (1961) that the $[X_2^-]^+$ center had a $\langle 110 \rangle$ symmetry axis; so, Kabler and Murray and Keller knew that all their $[X_2^-]^+$ centers were along one of these $\langle 110 \rangle$ axes. Finally, they released the electrons which had been trapped at the impurities by optical bleaching in the infrared and measured the polarization of the resulting exciton luminescence. This was polarized either parallel or perpendicular to the axis of the $[X_2^-]^+$ center and showed that the exciton, too, had a $\langle 110 \rangle$ symmetry axis. Their discoveries subsequently influenced the formulation of the mechanism of photochemical radiation damage in alkali halides (Pooley, 1965; Hersh, 1966).

5. VIBRONIC STRUCTURE AND ZERO-PHONON LINES

So far, we have not considered the way the electron that is excited in an optical transition may also interact with lattice vibrations or with local modes of vibration around the defects. It is, of course, this interaction that gives the width to broad bands like the F band, where the width varies with temperature according to the relation (Markham, 1959)

$$W^2(T) = W^2(0) \coth (hw/2kT). \tag{7}$$

$W(0)$ is the low temperature width of the band and hw is the energy of the quantum of the single local mode assumed to be interacting with the electron. Although it is known that many vibrational modes interact with most defects, it is often a good approximation to assume that only one vibrational mode dominates the interaction. One justification for this is the good fit to the equation for $W(T)$ given above of measurements made on the F center in KCl and KBr (Gebhardt and Kuhnert, 1964). In the case of the F center, the mode that dominates is the symmetric, breathing mode of the center, in which the six cations surrounding the F center vacancy move inward and outward together. A more elaborate approach has been provided by Möstoller et al. (1971), which takes into account an averaging of the lattice modes interacting with the defect.

Some defects have optical absorption spectra much narrower than those of F centers, and with well-defined vibronic structure (Fig. 8). Figure 8 also

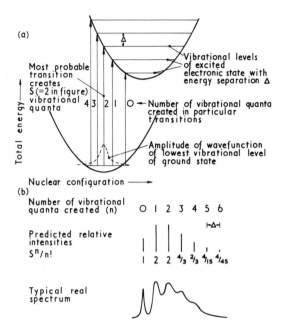

(a)

Total energy →

Nuclear configuration →

(b)

Fig. 8. Vibronic structure and the configuration coordinate model; (a) a configuration coordinate diagram, showing the variation of energy with nuclear configuration for two electronic states and the optical transitions between the states; (b) the optical absorption spectrum predicted by the configuration coordinate model and a typical real spectrum.

illustrates the configuration coordinate model, which gives a simple account of this vibronic structure. The basis of the model is the Born–Oppenheimer approximation of quantum mechanics, which assumes that the electrons of a solid or molecule are able quickly to adopt the lowest energy configuration of given symmetry appropriate to a given nuclear configuration. The total energy of each electronic state will therefore depend on the nuclear configuration (Fig. 8a), having a minimum at the equilibrium configuration. Usually, the configuration of minimum energy in an excited state is different from that for the ground state, with the result that the most probable optical transition, the one directly upward in Fig. 9a from the configuration where the ground vibrational state wave function is a maximum, involves creating several (say S) vibrational quanta. This does assume that only the lowest vibrational level of the ground electronic state is populated, which is nearly always valid at low temperatures. It is then not difficult to show (e.g., Fitchen, 1968) that the probability of a transition in which n vibrational quanta are created is

$$\rho_n = S^n \exp{(-S)}/n! \tag{8}$$

Although the configuration coordinate model implies *strong* coupling to a *single* vibrational mode, the expression for ρ_n is the same for *weak* coupling to a *large number* of equivalent modes. This latter situation was treated by Huang and Rhys (1950) and for this reason the coupling parameter S is usually called the Huang–Rhys factor. In this case, S represents the number of modes that are excited by a single quanta rather than the number of quanta by which a single mode is excited.

Note that the probability of the zero-phonon transition ($n = 0$) is just $\exp(-S)$. For the F center in alkali halides, values of S are 20–40 (e.g., Dawson and Pooley, 1969) so that the zero-phonon line can never be seen. Many F aggregate centers, however, have much smaller values of S and show spectra as at the bottom of Fig. 8b, where the zero-phonon line is clearly visible. As shown in Fig. 8b, the phonon-assisted transitions are always much broader than the zero-phonon lines; this is to be expected since the electron does not actually interact either with a single vibrational mode or a large number of equivalent modes, but a number of different modes, making one and multiphonon transitions of a variety of energies possible. The residual width of zero-phonon lines is determined largely by lattice strains (Stoneham, 1966; Hughes, 1968).

6. PERTURBATION EFFECTS

Perturbation effects on both phonon lines and broad bands have been used in defect structure analyses. Uniaxial stress is particularly effective in these studies, but electric and magnetic fields are also used.

Defects that give rise to zero-phonon lines constitute the optimal system for the observation of uniaxial stress effects. The electron–lattice interaction is strong enough for the stress, which affects the lattice, to change significantly the energies of the electronic transition, and yet weak enough for the line width of the zero-phonon line to be small and its intensity large. The result is that, when uniaxial stress is applied to a crystal, zero-phonon lines often move significantly and split into several components. The splitting may occur because of the removal of orientational or electronic degeneracy and can best be described group theoretically (Kaplyanski, 1964; Hughes and Runciman, 1965; Fitchen, 1968). It is easy, however, to illustrate the physical basis of the effect of stress with simple examples. M-like centers have $\langle 110 \rangle$ orientational degeneracy; each can have six possible orientations, and in an unstressed crystal all are equivalent (see Fig. 7a). Centers with $\langle 110 \rangle$ symmetry do not have associated electronic degeneracies (Runciman, 1965). If stress is applied along [100], then the centers along [110], [1$\bar{1}$0], [10$\bar{1}$], and [$\bar{1}$0$\bar{1}$] will all be affected in the same way since they are all at

45° to the applied stress, whereas the centers on [01$\bar{1}$] and [0$\bar{1}\bar{1}$] are at 90° and will be affected differently. The zero phonon would then split into two components, with relative strengths depending on the direction and polarization of the measuring light. Similarly, a $\langle 110 \rangle$ stress would separate the six orientations into three groups, one center parallel to the stress, one perpendicular to it, and four at 60°. A $\langle 111 \rangle$ stress would separate the centers in two groups, three centers at 90° to the stress and three at arccos $(\sqrt{2}/\sqrt{3})$. Unfortunately, the zero-phonon lines of M centers in alkali halides are always too weak for observation, but the M$^-$ center in LiF does have an adequately strong zero-phonon line and has been studied (Fitchen et al., 1966).

The removal of electronic degeneracy by stress is similar. Suppose we had a cubic center such as the F center in which the zero-phonon line could be seen. For the F center, the electronic transition is largely S $\rightarrow$ P and therefore triply degenerate, but removal of the cubic symmetry by applying a $\langle 100 \rangle$ stress would split the p state into two components, one nondegenerate and one doubly degenerate. For halide F centers, zero-phonon lines are not observable, but the F$^+$ center in CaO does behave in this way (Hughes and Henderson, 1972).

Direct Zeeman splitting of defect zero-phonon lines has never been seen, even at very high pulsed fields in defects known to contain odd numbers of electrons (Hughes and Runciman, 1965). Hughes and Runciman attributed the absence of a splitting of the R$_2$ line in LiF to the fact that both ground and excited states had g values near 2 and that only $\Delta M_s = 0$ transitions were allowed. Even when simple Zeeman splittings are too small to be observable, however, magnetic circular dichroism or Faraday rotation effects in paramagnetic centers can often be detected. In this case, the equilibrium populations of the magnetically split ground state become significantly different at low temperatures. Since absorption of left and right circularly polarized light corresponds to excitation from the two spin components of the ground state, the real and imaginary parts of the dielectric constant for the two polarizations differ. Circular dichroism (Karlov et al., 1963) measures the imaginary part, that is, the difference in absorption coefficients, while Faraday rotation (Lüty and Mort, 1964) measures the real part, the difference in refractive index. These techniques have been particularly valuable in studies of F centers in halides and F$^+$ centers in oxides.

It should now be clear that optical techniques are extremely powerful in the study of the physics of color centers and of radiation damage processes, since they (i) aid in giving a complete determination of the structure of defects and their interaction with the lattice, (ii) allow a very fast and fairly selective measure of defect concentrations with which monitoring can be carried out both during and after irradiation.

D. Magnetic and Electric Properties of Color Centers

It has been mentioned that magneto-optic effects can help in defect structure determination. In this section we will therefore consider only the wholly magnetic and electrical effects exhibited by color centers, such as magnetic susceptibility. As one simple example, it is possible to distinguish between paramagnetic centers like halide F centers and diamagnetic centers like the F_2 centers solely on the basis of magnetic susceptibility. Jensen (1939) helped to confirm the structure of the F center and Sonder (1962) that of the F_2 center in this way. However, simple magnetic susceptibility is far less powerful than paramagnetic resonance techniques such as EPR (Electron Paramagnetic Resonance) and ENDOR (Electron Nuclear Double Resonance) (Seidel and Wolf, 1968; Schumacher, 1970).

1. EPR IN STRUCTURE ANALYSIS

Simple EPR consists of exciting transitions between different spin states which have been split in energy by an applied magnetic field. Many defects, such as halide F centers, have a single unpaired electron in an orbitally nondegenerate state, and provide good examples of simple EPR. The mere observation of an EPR spectrum due to a defect indicates that it has at least one unpaired electron, but much more information is found through *g values* and *hyperfine interactions* (Abragam and Bleaney, 1970).

The splitting by a magnetic field H of a spin doublet state is usually written as

$$\Delta E = g\beta H \qquad (9)$$

where β is the Bohr magneton and g is called the spectroscopic splitting factor, or g value. For most defects, g is close to the free electron value of 2.0023, so that a field of 0.3 T gives a splitting of 0.3 cm, corresponding to a resonance frequency of ~ 9 GHz. In fact, the value of g usually differs slightly for different directions of H because of spin orbit interaction. In order to account for this the splitting energy is more correctly written $\beta \mathbf{S} \cdot g \cdot \mathbf{H}$ where $\mathbf{H}$ is now the magnetic field vector, $\mathbf{S}$ the electron spin vector, and g is a tensor. The g tensor has the same symmetry as the defect and it is possible to determine the latter from the former, thus gaining useful information about defect structure. In CaO, where there are no abundant magnetic nuclei, only g-value information is available from EPR measurements, and g-value analyses of many defects in CaO have been made by Bessent (1969).

Most crystals are not like CaO, they contain atoms that have magnetic nuclei; for example ^{25}Mg has a spin of $\frac{5}{2}$. In these crystals the spins of unpaired electrons interact with the nuclear spins of nearby magnetic nuclei and give hyperfine interaction. Figure 9 shows the derivative with respect to magnetic field of the EPR absorption of the F^+ center in MgO (Wertz *et al.*, 1957). In this case some of the F^+ centers have only nonmagnetic ^{24}Mg and ^{26}Mg as nearest neighbors; the next nearest neighbor oxygen ions are almost always nonmagnetic since the only stable oxygen isotope that has a magnetic moment is ^{17}O, and this occurs only in very low concentration (0.037%). The F^+ centers cause the strong central line. Some of the F^+ centers have one ^{25}Mg as nearest neighbor, and interaction between the

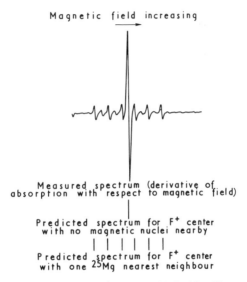

Fig. 9. The EPR spectrum of the F^+ center in MgO. After Wertz *et al.* (1957).

electron and its nuclear spin changes the electron energy levels through *hyperfine interaction*. The interaction energy is

$$\Delta E_h = \mathbf{S} \cdot A \cdot \mathbf{I} \tag{10}$$

where $\mathbf{S}$ is again the electron spin vector, $\mathbf{I}$ the nuclear spin vector, and A is the hyperfine tensor. Since $I = \frac{5}{2}$ for ^{25}Mg each electron spin level is split into six components (Fig. 10). Only transitions for which $\Delta M_I = 0$ are allowed, so the main EPR transition is split into six components of equal intensity for these F^+ centers. A smaller number of centers have two ^{25}Mg ions in the nearest neighbor shell of six. In suitable configurations these two ions can be equivalent, and in this case the line is split into eleven components with relative intensities $1 : 2 : 3 : 4 : 5 : 6 : 5 : 4 : 3 : 2 : 1$. Thus hyperfine interaction can give quite detailed information about the structure of the center, as well as giving the symmetry in a way analogous to that for the g

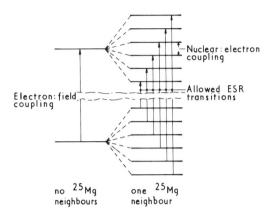

Fig. 10. Hyperfine splitting of the energy levels of F^+ centers in MgO in a magnetic field.

value. It is even possible to determine the extent of the unpaired electron wave function from the hyperfine interaction, by using the fact that the isotropic part of the A tensor depends on the probability of finding the unpaired electron in the very near vicinity of the nucleus with which it is interacting.

One of the most notable uses of EPR in the structural analysis of defects was made by Castner and Kanzig (1957) in their work on the $[X_2^-]^+$ center. Figure 11 shows the EPR spectrum of $[F_2^-]^+$ centers in LiF. The dominant three line spectra arise from the interaction between the unpaired electron and two equivalent ^{19}F nuclei, each having spin $\frac{1}{2}$. Taken as a whole the EPR spectrum gives convincing evidence for the now accepted model for the $[X_2^-]^+$ center. Other hole centers have been dealt with in the same way (Seidel and Wolf, 1968).

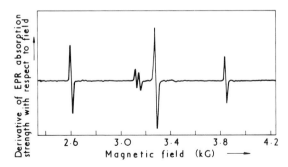

Fig. 11. The EPR spectrum of $[F_2^-]^+$ centers in LiF, measured at 77 K with the magnetic field along a $\langle 100 \rangle$ axis. The spectrum shows a three line structure with a $1 : 2 : 1$ intensity ratio characteristic of strong interaction with two equivalent spin $\frac{1}{2}$-(^{19}F) nuclei. After Castner and Kanzig (1957).

2. EPR in Radiation Damage Studies

As in optical spectroscopy the total intensity of an EPR absorption spectrum is proportional to the number of centers that are absorbing. There is a difference, namely that nothing equivalent to the oscillator strength for optical spectra prevents the strength of the EPR signal giving an absolute value for the number of absorbing centers. In fact absolute measurements of EPR signal strengths are difficult, but they can be made, and are even used to determine the oscillator strengths of optical transitions.

The use of EPR and optical absorption simultaneously has led to the identification of the F center optical band in MgO in a very satisfying way. First Henderson and King (1966) were able to show that the optical band at 4.95 eV, produced by neutron irradiation of MgO, correlated with the EPR signal of the F^+ center. They therefore assigned the 4.95-eV band to the F^+ center. Subsequently Chen et al. (1968) showed that the correlation broke down in additively colored or electron irradiated crystals, and were able to confirm that this happened because part of the optical band was due to diamagnetic F centers, which also absorb at around 5 eV.

The final and most convincing evidence that the F center production mechanism in alkali haldies involves electron–hole recombination was provided by the EPR experiments of Keller and Patten (1969). The key factor in their approach was the ability of EPR to distinguish clearly between H and $[X_2^-]^+$ centers, which is not possible optically. They also used the different thermal stabilities of these two centers in a very elegant way. Using KCl crystals containing electron trapping impurities they first X irradiated at ~ 77 K. At this temperature the major products of irradiation are F centers, $[X_2^-]^+$ centers and impurities which have trapped electrons; no H centers remain. They then cooled their crystals to ~ 4 K and released electrons from the impurities by infrared irradiation. The electrons recombined with the $[X_2^-]^+$ centers, the disappearance of which they could monitor with the EPR spectrum, and in doing so created H centers. They found an exact correlation between disappearance of the $[X_2^-]^+$ centers and the formation of the H centers, which is very direct evidence for the reaction

$$[X_2^-]^+ + e \rightarrow F + H \tag{11}$$

A similar experiment using optical absorption would be much more equivocal if not impossible.

3. Endor

It often happens that the width and number of hyperfine components of EPR lines are too great for resolved hyperfine structure to be seen. An example of this is the F center in KCl, where the F center electron interacts with a large number of magnetic nuclei and which therefore shows a single broad EPR line with no resolved components. Even in more favorable cases it is rarely possible to resolve the interaction between the free electron and nuclei in shells beyond the nearest and next nearest neighbors (Seidel and Wolf, 1968, Table 8-3). ENDOR techniques have much better resolution and can be used under these conditions.

The basis of the ENDOR effect (Feher, 1959; Seidel and Wolf, 1968) can be illustrated by Fig. 12, which shows the interaction of an electron and a nuclear spin of $\frac{1}{2}$ with a magnetic field and with each other. The allowed EPR transitions ($\mid -+\rangle$ to $\mid ++\rangle$ and $\mid --\rangle$ to $\mid +-\rangle$) in the Dirac notation are both partially saturated by applying a fairly high microwave power at the EPR frequency. Although the two EPR transitions overlap, if

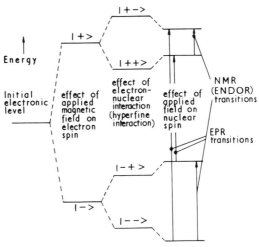

Fig. 12. Energy levels and EPR and ENDOR transitions for an electron coupled to a single nucleus of spin $\frac{1}{2}$ in a magnetic field.

they do not, then there is no need to use the ENDOR technique, in general one will be saturated to a different extent than the other. If a second radio frequency is introduced, having the right energy to cause the nuclear resonance transitions $\mid --\rangle$ to $\mid -+\rangle$ or $\mid ++\rangle$ to $\mid +-\rangle$, then the two EPR transitions will be mixed and the total microwave absorption of the partially saturated specimen will change, so that the positions of the nuclear resonance lines can be determined. Whereas, the line widths of EPR lines are usually 5–100 MHz for defects, the widths of the ENDOR lines are typically those of nuclear magnetic resonance (NMR) transitions (10–100 kHz), a fact that gives ENDOR its greatly improved resolution. On the other hand, the sensitivity of ENDOR is essentially that of EPR rather than NMR spectroscopy. Figure 13 shows the ENDOR spectrum of F centers in KCl (after

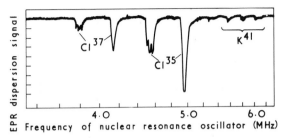

Fig. 13. The ENDOR spectrum of F centers in KCl, measured at 1.2 K with an EPR frequency of $\sim$9 GHz. After Feher (1957).

Feher, 1957). Compared with the EPR spectrum, a single gaussian line, the increase in resolution is dramatic.

4. Ionic Conductivity, Dielectric Loss, and Ionic Thermocurrents

Defects in ionic crystals are frequently electrically charged, and their motion causes ionic electrical conductivity. In simple insulators the conductivity is usually large enough to measure only at relatively high temperatures because the activation energy for defect motion is fairly high (often 1–2 eV). Nevertheless, at these temperatures ionic conductivity is a significant source of information (Fuller, 1972).

In alkali halides the defect mainly responsible for electrical conductivity is the positive ion vacancy. This exists in significant numbers even in pure crystals if the temperature is sufficiently high because at high temperatures the equilibrium concentration of Schottky defects (pairs of positive and negative ion vacancies) is large. However, the concentration of positive-ion vacancies can be considerably enhanced by the addition of divalent cations, which are often charge compensated in alkali halides by these vacancies. For example, Fig. 14 shows how the addition of $SrCl_2$ to KCl causes an increase in the positive-ion vacancy concentration and hence in ionic conductivity (Chandra and Rolfe, 1970). The slope of the graph of log σT versus $1/T$ for the impure specimens gives essentially the activation energy for the *motion* of the positive ion vacancies. On the other hand, the slope for the pure crystal has an activation energy composed of that of *motion* and of *formation* of the positive and negative vacancy pairs.

In equilibrium at low temperatures divalent cations are bound to their charge compensating vacancies, forming electrically dipolar defects. These, and most other defects that are electric dipoles, can be preferentially aligned in an electric field. If an alternating electrical field is applied and its frequency varied, resonance peaks can be found at which the dielectric loss of the dipole containing sample passes through a maximum (Nowick, 1972).

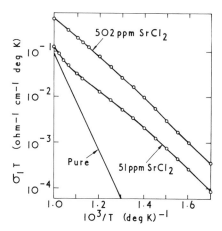

Fig. 14. The ionic electrical conductivity of KCl : $SrCl_2$ crystals as a function of temperature. After Chandra and Rolfe (1970).

This is because at frequencies above that of the resonance peak the dipoles do not have time to align themselves at all in the field and no polarization occurs; at frequencies below resonance the polarization is able exactly to follow the voltage, so that the displacement current is 90° out of phase with the voltage and no power loss is incurred. The frequency at which a dielectric loss peak occurs (ω) is related to the relaxation time (τ) for the dipolar defects by $\omega\tau \sim 1$.

The magnitude of the dielectric loss peak is proportional to dipole concentration, and this feature has been used by Cooke and Dryden (1962) to study the way divalent–cation, cation–vacancy pairs in NaCl and KCl aggregate at room temperature. They were able to show that these dipoles rather surprisingly aggregated into trimers.

It is usually easier not to measure the dielectric loss peak by changing ω at constant temperature, but instead to work at constant ω and change τ by changing the temperature (Fig. 15). The relaxation time varies with temperature roughly according to the relation

$$\tau \sim \tau_0 \exp\left(E/kT\right) \tag{12}$$

where E is the jump energy of the defect so that relatively small changes in temperature can often be equivalent to large changes in frequency. Dielectric loss measurements have a potential sensitivity of about 10^{23} dipoles m^{-3}.

Another, more sensitive technique for detecting dipolar defects is by ionic thermocurrents or ITC (Bucci and Fieschi, 1964). The crystals are first polarized by applying a steady electric field for a time long compared with the relaxation times at the temperature used. They are then cooled, with the field

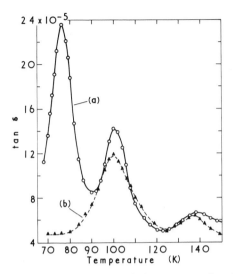

Fig. 15. The dielectric loss of a synthetic crystal of α-quartz as a function of temperature at 2 kHz: (a) with the electric field parallel to the z axis; (b) with the electric field perpendicular to the z axis. After Nowick (1972).

still applied, to a temperature where the relaxation times for all the dipoles in the crystal are long, i.e., the polarization is "frozen in." The applied field is then removed and replaced with a sensitive current detector and the crystal is then warmed at a steady rate. As the temperature rises over a particular temperature regime, the defects relax from their preferentially aligned condition and a depolarization current flows. Thus a particular defect will give an ITC current peak at a characteristic temperature (the peak temperature depends only slightly on the heating rate). From the shape, position, and area of the ITC peak it is possible to determine E, τ_0, and the number of defects contributing to the peak (Fig. 16).

Using ITC measurements Stott and Crawford (1971) have identified dipolar defects, produced by X irradiation of alkali halides, in numbers comparable with the F centers produced. They were led to postulate that the irradiation must be producing defects in the cation as well as in the anion sublattice, a surprising and somewhat contentious suggestion. However, the electron microscopy of interstitial clusters carried out by Hobbs *et al.* (1972a,b, 1973) provides indirect evidence for their point of view from quite different measurements on the fate of interstitial atoms in alkali halides. This particular point illustrates very well the need for application of a wide variety of techniques to a problem since neither optical nor magnetic resonance measurements have provided any information about the dipolar defects "discovered" by ITC and deduced from electron microscopy.

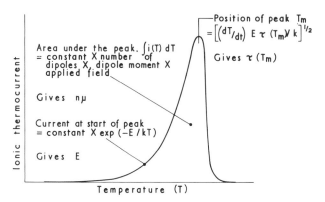

Fig. 16. A typical ionic thermocurrent peak, showing the information available in different features of the curve. The rate of temperature rise is dT/dt. The symbol μ denotes the dipole moment.

E. Mechanical and Thermal Properties of Color Centers

Mechanical and thermal properties can also be changed significantly by the presence of defects, and these changes provide further techniques in defect studies, sometimes giving information unobtainable by other means. It must be admitted, however, that the wealth of information available about the microscopic structure of point defects in optical and magnetic resonance spectroscopy has no analogy in that derived from thermal and mechanical properties.

1. VOLUME AND LATTICE PARAMETER CHANGES

The volume changes that accompany color center production in alkali halides have been measured by a variety of techniques. Direct measurement, by flotation of X-rayed crystals, were used by Estermann *et al.* (1949) and have been used more recently in the study of oxides that have been neutron irradiated (Henderson and Bowen, 1971). In all cases the volume of irradiated crystals is more than before irradiation and their density less. Although clever refinements to the direct flotation method of measuring density have been made (Witt, 1952), the most sophisticated and useful technique for measuring changes in volume is undoubtedly via electronic techniques (e.g., Farnum and Royce, 1968). In this instance one end of the crystal being irradiated is fixed and the other is made one plate of a capacitor. A second, fixed, plate of the capacitor is placed near the first, so that small fractional changes in the length of the crystal cause much larger changes in capacitance, which can then be detected electronically. Apart from the considerable advantage of high sensitivity this technique is preferred because measurements can easily be made during irradiation and at low temperatures.

In themselves radiation-induced volume changes do not indicate much more than that defects are formed in crystals by irradiation. In conjunction with measurements of lattice parameters, however, they have been the major tool in deciding whether irradiation of alkali halides produced only vacancies or vacancies plus interstitials. If only vacancies are produced, we expect the volume of a crystal to grow since new unit cells are introduced; but we do not expect the average lattice parameter to change. The latter prediction is based on the smallness of the relaxation of ions around vacancies; intuitively we might expect the ions to relax inward a little, which would constitute a reduction in average lattice parameter. On the other hand, if interstitials are produced as well as vacancies, then no new unit cells are introduced but the interstitials will cause a significant increase in average lattice parameter, since the relaxation of the ions around interstitials is likely always to be large and outwards. Theoretical treatments show (e.g., Eshelby, 1953, 1954) that for Frenkel defects, that is pairs of interstitials and vacancies, the fractional change in lattice parameter should be equal to the fractional change in macroscopic length. For Schottky defects, the lattice parameter change is much smaller than the length change.

This analysis, when applied to alkali halides, soon showed that Frenkel defects were formed by irradiation (Fig. 17) (Binder and Sturm, 1954). Very refined measurements by Peisl and his co-workers (Peisl *et al.*, 1966; Balzer *et al.*, 1966), have confirmed Frenkel defect production in a number of alkali

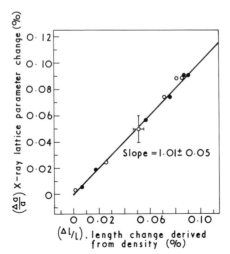

Fig. 17. Comparison of length and lattice parameter changes in LiF caused by neutron irradiation. After Binder and Sturm (1954).

halides, at low temperatures as well as at room temperature, and similar measurements have been made on oxides (Henderson and Bowen, 1971).

2. X-Ray and Neutron Scattering

When the lattice parameter of a crystal containing defects is measured, the effect of the defects is apparent in the shift of the main Bragg diffraction peak. This does not give any direct information about the structure of the defect. However, besides causing changes in lattice parameter throughout the whole crystal, defects also cause more severe distortion and bending of the lattice planes in their vicinity. The effect of this local distortion is to produce wings on the main Bragg diffraction peak; these can be used to obtain information about the strength and symmetry of the defect strain fields.

The technique used by Trinkhaus *et al.* (1970) is to measure very carefully the shape of the coherent X-ray diffraction peak, keeping the thermal-diffuse-scattering intensity small by cooling the crystals to 80 K. Since expected defects cause wings to appear, the overall shape can be compared with that predicted by a variety of theoretical models. In this way Spalt and Peisl (1971) showed that interstitial clusters containing 200 atoms were formed in X-irradiated LiF.

A very different technique was developed by Martin (1962; Martin and Henson, 1964); that of cold neutron scattering. Coherent scattering of X rays or neutrons is not possible at all if their wavelengths are greater than $2d_{max}$, where d_{max} is the maximum interplane separation on the crystal. A perfect crystal should therefore be transparent to such particles provided no inelastic processes occur, but a crystal containing defects should scatter the particles. Because of their high energy, X rays can undergo many inelastic collisions which involve excitation of electrons even in perfect crystals, whereas neutrons are only able to excite certain types of lattice vibration and are scattered strongly only by defects in crystals. "Cold" neutrons, produced by thermalizing neutrons in a liquid hydrogen moderator and therefore with long wavelengths, are very suitable for use in this way and were chosen by Martin. He was able to fit his scattering data on irradiated graphite to a theoretical model and show that the clusters formed contained only five atoms. Clusters of this size are not normally observable, either with the techniques for point defects we have discussed in this chapter or with electron microscopy, so that cold neutron scattering can be an important tool. It has not been widely used; because large concentrations of defects are necessary, specimens must be rather large and unequivocal interpretation is very difficult. Experiments have been carried out by Martin (1968) on MgO and Clarke and co-workers (1971) on germanium.

3. Electron Microscopy

The mechanical changes caused by defects in crystals are obviously largest near the defects; it is near the defects that ions are most disturbed from their proper lattice sites. Often these mechanical changes can be detected directly with electron microscopy. A thin crystal foil is usually placed in a monoenergetic and uniform intensity electron beam at an angle near that appropriate for Bragg scattering of the electrons. The electrons transmitted (or diffracted) are imaged onto a photographic film or some other detector. Near defects the lattice planes are bent either nearer to or further from the Bragg condition. In either case different numbers of electrons are scattered and intensity contrast appears in both transmitted and diffracted images. The resolution of the image is usually good enough to permit the direct viewing of defects which are large enough ($> 10nm$).

It follows that electron microscopy is a very powerful technique for studying large defect clusters and has been used effectively in studies of dislocations (Bowen and Clarke, 1964) and voids (Morgan and Bowen, 1967) in MgO. It is not possible in this chapter to describe adequately the amount of information about large clusters that can be derived using the electron microscope; instead we will mention one point which arises out of the recent work of Hobbs et al. (1972a,b, 1973).

Hobbs (1970) developed a technique that overcame the difficulties of preparing and using thin alkali halide films for microscopy. This made it possible to view directly the interstitial clusters formed in parallel with F centers in room temperature irradiated alkali halides (Fig. 18). This in itself is a significant step forward since the structure of the clusters is much more apparent in the electron microscope than through any of the indirect methods used previously such as flow stress, thermal conductivity, or diffuse X-ray scattering. However, the dislocation loops produced, at least in KI (Fig. 18), appear to be perfect unfaulted dislocation loops composed of equal numbers of anions and cations. This is very surprising since all the theories of radiation damage in alkali halides predict that anions only will be displaced, and no hard evidence for cation defects exists. The present thinking about this is that halogen interstitial atoms start to cluster but in doing so cause the formation of cation–anion vacancy pairs containing halogen molecules together with perfect dislocation loop. The postulate of the vacancy pairs is in accord with the discoveries by ITC of Stott and Crawford (1971).

4. Flow Stress Measurements

The flow stress of a crystal is determined experimentally by obtaining the stress–strain curve for a given specimen, as shown in Fig. 19 for MgO crystals (McGowan and Sibley, 1969), and defining the point where there is a

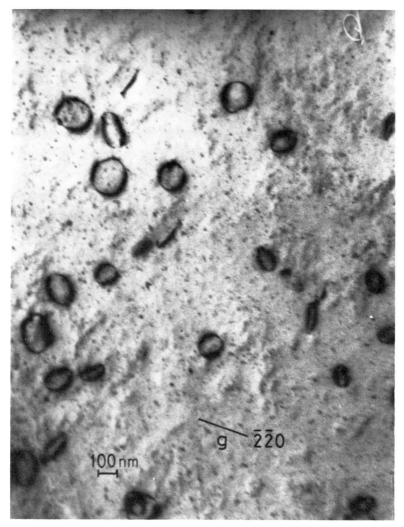

Fig. 18. Large defects, interstitial clusters in alkali halides, seen in the electron microscope. After Hobbs *et al.* (1973).

deviation from Hooke's law as the flow stress. From the figure it is evident that irradiation of MgO with neutrons or electrons increases its flow stress, and it is already well known that even gamma irradiation can increase the flow stress of alkali halide samples (Crawford, 1968). This suggests that defects introduced by irradiation are in some way active in hardening crystals. Since the deformation or flow of crystals involves dislocations moving on certain lattice planes (Cottrell, 1953; Friedel, 1963), it is very probable

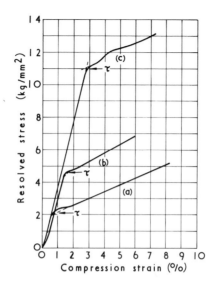

Fig. 19. Stress–strain curves for unirradiated and irradiated MgO crystals: (a) as received crystal; (b) crystal irradiated with 1.2×10^{18} electrons/cm^2 (1.8 MeV); (c) crystal irradiated with 8.3×10^{16} neutrons/cm^2 (>1 MeV). The flow stress is shown by the arrows. After McGowan and Sibley (1969).

that the interaction between these dislocations and point defects causes the hardness of the crystals to increase on irradiation. On this basis one would imagine that adding impurities to crystals would also increase their hardness. This is indeed observed in many cases. In fact, any obstacle, impurity, radiation defect, or dislocation dipole that produces an elastic strain field in the crystal lattice will interact with moving dislocations in such a way as to increase the flow stress. Moreover, since in ionic crystals dislocations can be charged, it is also possible for electrostatic interactions between charged defects and charged dislocations to harden these crystals.

Fleischer (1962a,b) has developed a theory for the hardening of crystals produced by defects. He discusses the effect on the flow stress (τ) of divalent impurity–vacancy pairs and radiation induced defects that produce tetragonal strain fields. His relationship between the stress required to move a screw dislocation and the strain field of tetragonal defects is

$$\tau = (Gc^{1/2}/3.3)\, \Delta\varepsilon \tag{13}$$

where $\Delta\varepsilon$ is a measure of the tetragonality of the defect, G is the shear modulus, and c is the impurity concentration in parts per million. The tetragonality of a defect $\Delta\varepsilon$ is defined as the fractional difference between the strains parallel to, and perpendicular to, the axis of the defect. If this theory is fitted to experiments with impurities in the alkali halides, a value for $\Delta\varepsilon$ of about 0.5 is necessary. For impurities, there is a strong temperature dependence of flow stress on impurity concentration since the impurities can precipitate, form dimers, trimers, etc., and all of these possibilities are very temperature dependent. In addition, at temperatures sufficiently high for impurity–vacancy pairs reorientation, Pratt and his co-workers (1963) point out that as the dislocation moves toward this type of defect the stress field of the dislocation will tend to reorient the dipole. This effect is called the Snoek effect and can be very important in metal crystals as well as insulators.

The Fleischer theory predicts that F centers should have little effect on the flow stress since they are symmetric defects and have $\Delta\varepsilon \simeq 0$. However, isolated interstitials should give a marked increase in flow stress when they are introduced into the crystal. Figure 20 (Nadeau, 1963) does show the

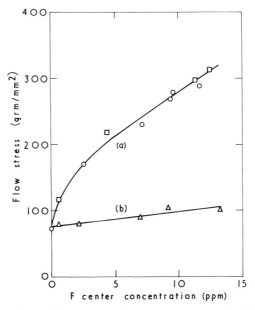

Fig. 20. The dependence of flow stress on F center concentration in (a) irradiated, and (b) additively colored KCl. After Nadeau (1963).

marked difference between additively colored crystals, containing only F centers, and irradiated crystals containing both interstitials and vacancies. However, since the interstitials are not present in isolated form, the Fleischer theory cannot be applied directly. Nevertheless, Nadeau (1964) has made a study of radiation hardening in 13 of the 20 alkali halides and concluded that in all cases the hardening defect is an interstitial halogen resulting from Frenkel pair production. McGowan and Sibley (1969) have shown that interstitials are responsible for the hardening of electron-irradiated MgO.

As mentioned previously, it is extremely important when dealing with the complex problems of defects in insulators to be able to use as many experimental tools as possible. In the case of stress measurements it is possible not only to investigate the hardening of materials by impurity doping or irradiation, but also, once the hardening mechanisms have been determined, it is useful to monitor the distribution of the hardening defects by means of flow stress measurements. In other words, we are able to use dislocations to probe the crystal lattice and determine the distribution of interstitials produced by the radiation and the state of impurities at various temperatures.

5. THERMAL PROPERTIES

In principle the introduction of defects changes both the specific heat and the thermal conductivity of crystals. In practice neither is significantly changed by the concentration of defects normally achievable in ionic solids at ambient temperatures and only thermal conductivity at low temperatures has proved a useful tool.

Walker and Pohl (1963) studied the low temperature thermal conductivity of KCl containing about 0.1% KI. As expected the conductivity was depressed, but it was depressed particularly near 20 K, where a pronounced dip occurred. Elliott and Taylor (1964) were later able to give a good theoretical account of the general shape of the curve and of the dip, which occurred at temperatures where the frequencies of the phonons most involved in conductivity corresponded with a resonance vibration of the impurity ion.

Sonder and Walton (1967) made use of thermal conductivity changes at very low temperatures to sense the presence of interstitial clusters (Fig. 21). By relating the temperature at which the plot of log thermal conductivity versus log temperature changes its slope to the wavelength of the phonons being scattered, and hence to the size of the scattering defect, they showed that clusters of ~ 1000 atoms were produced by room temperature irradiation of KCl. Although this observation has now been reduced in significance by the harder evidence for clusters of diffuse X-ray scattering and electron microscopy, at the time it provided the only unequivocal evidence for clusters and information about their size which was available.

In Section II we have looked at many of the ways in which defects alter the physical properties of ionic solids, and tried to show how these changes can be used to provide information about defect structure and the radiation damage process. We are now able to look in detail at the processes involved in radiation damage.

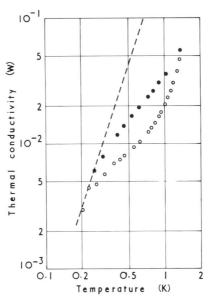

Fig. 21. Thermal conductivity of KCl crystals below 2 K; the dashed line represents pure KCl, filled circles a crystal irradiated at ~ 80 K to an F center concentration of 6×10^{18} cm^{-3} and open circles a crystal irradiated at ~ 300 K to an F center concentration of 3×10^{18} cm^{-3}. After Sonder and Walton (1967).

III. Radiation Damage in Insulators

A. *The Need for High Quality "Pure" and Impurity-Doped Crystals*

As mentioned in the introduction, radiation damage mechanisms can be exceedingly sensitive to the presence of impurity ions in the crystal matrix (Schulman and Compton, 1962; Crawford, 1968; Sonder and Sibley, 1972). Furthermore, it often happens that impurities have optical transitions in the very region of the spectrum where radiation-produced defects also have absorption bands. This interference or overlap of the optical absorption bands can preclude the use of optical techniques for studying radiation damage. For example, in MgO it was not really possible to study vacancy-interstitial (Frenkel pair) production until the Fe^{3+} impurity concentration was reduced to less than 10 ppm since the Fe^{3+} absorption bands overlapped with the F^+- and F-center bands.

It is only within the last decade that high quality "pure" crystals (impurity concentration less than 5 ppm) of either the alkali halides or alkaline earth oxides have become available. Numerous groups have now produced excellent alkali halide crystals and good quality alkaline earth materials have been grown by W. & C. Spicer Ltd. (Henderson and Wertz, 1968) and Oak Ridge National Laboratory (Butler et al., 1971; Abraham et al., 1971). The quality of the alkaline earth halide crystals available is improving but is probably not yet as good as that of the alkali halides (Guggenheim, 1961; Nassau, 1961).

There are two important aspects of the production of high purity crystals. First, it is necessary to prepare very pure starting material and to grow the ingots in such a way that the starting material is not contaminated by the system but is purified even further. Secondly, and equally important, is the characterization of the ingot in terms of mechanical and chemical purity. This is done preferably by nondestructive techniques such as optical absorption or emission, or electron paramagnetic resonance; or with semidestructive techniques such as chemical etching. However, it is usually necessary, initially, to use destructive techniques such as wet chemistry, activation analysis, flame photometry, and flow stress measurements to characterize parts of the ingots.

The growth of very pure alkali halide and alkaline earth halide crystals can be accomplished by either zone-refining, Stockbarger, or Kyropolus methods (Peech et al., 1967; Grundig, 1965; Warren, 1965; Butler et al., 1966). In any case pretreatment of the starting powder is required to reduce OH^- contamination as well as to eliminate some of the other major impurities. At the present time, it is possible to produce pure alkali halide crystals with such small impurity concentrations that a quantitative measure of the impurity concentration is not possible. Table II illustrates the detection limits (in $\mu gm/gm$) for the methods generally used for quantitative chemical analysis (Butler et al., 1966; Willis, 1962), and Table III shows some typical analytical results ($\mu gm/gm$) for a few KCl crystals grown at Oak Ridge National Laboratory (Butler et al., 1966). It should be noted that in most cases the crystals are so pure that analysis merely indicates impurity levels below the limit of sensitivity of the technique.

Ionic conductivity is a very sensitive technique which can sometimes measure the *total* polyvalent impurity concentration of an ingot (Peech et al., 1967), but despite its great sensitivity, this method also has limitations. It is not selective in that it can rarely distinguish between different types of polyvalent impurities. More important, it does not always detect polyvalent impurities. In some cases chemical association of two impurities, such as Ca^{2+} and OH^- into $Ca(OH)_2$, produces an uncharged complex with no compensating effectively charged vacancies and despite the fact that polyvalent impurity ions are present, they go undetected in ionic conductivity measurements. It is possible, however, to reveal some of these molecular impurities with optical absorption measurements in the infrared region of the spectrum, once again pointing to the need for a multitooled approach, even in impurity analysis. Figure 22 portrays the $Ca(OH)_2$ absorption for a

TABLE II

Limits of Detectability (μgm/gm) for Various Analytical Techniques

Element or ion	Emission spectroscopy	Atomic absorption spectroscopy	Flame photometry	Wet chemistry	Neutron activation analysis	Optical absorption spectrometry
Ag		0.03			0.1	0.004
Al		1	1			
B	1					
Ba			0.5			
Be	0.01	300				
Bi		1				
Ca		0.1	1			
Cd	0.02	0.03				
Co		0.2	1		1	
Cr			0.1			
Cs			0.05		0.5	
Cu		0.06		1	0.02	
Eu			0.5			
Fe		0.2	0.2	0.1	0.5	
Hg		5		1		
In			0.1			
La			0.05			
Li		0.02	0.1			
Mg	0.6	0.01				
Mn		0.05	2		0.01	
Mo	4	0.4	2			
Na			1		0.1	
Ni		0.2				
Pb		0.5				0.004
Pt		1			0.8	
Rb		0.1	1			
Si				0.2		
Sn		10		10		
Sr		0.1	0.05			
Ti	0.3			3		
Tl		0.8	0.1			
V	4					
Zn		0.02				
Br				1	0.05	
I				1		
BO$_2$				0.2		0.01
PO$_4$				1		
NO$_3$				1		
CO$_2$				20		
OH						0.001

TABLE III

IMPURITY CONCENTRATIONS (μgm/gm) IN THREE
" PURE " KCl CRYSTALS[a]

Impurity	Crystal 1	Crystal 2	Crystal 3
Br	9.1	1.3	< 1
C	< 300	< 100	< 100
Fe	< 0.5	< 0.7	< 0.25
I	2.7	1.5	< 1
N	4	< 1	4
Na	2.9	< 0.6	< 0.5
OH	1.4	< 0.04	< 0.02
P	1.0	< 1	< 1
Pb	< 0.006	< 0.03	0.06
Rb	< 2	< 1	1
S	< 5	< 1	< 1

[a] Grown at Oak Ridge National Laboratory.

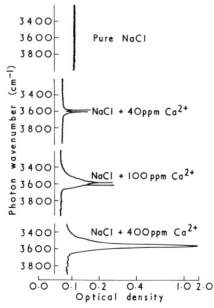

Fig. 22. Infrared absorption bands in calcium-doped NaCl crystals. The bands arise from OH stretching vibrations. After Johnston and Nadeau (1964).

series of crystals grown in air with the starting material doped with different amounts of $CaCl_2$ (Johnston and Nadeau, 1964). Notice that there is considerable impurity in these samples, as evidenced by the absorption, but since it is complexed, ionic conductivity measurements are insensitive to it.

Optical absorption or emission measurements are potentially excellent methods of nondestructively evaluating the impurity concentration in transparent materials. For example, Fig. 23 shows the Pb impurity absorption at 272nm versus the total Pb concentration in a KCl crystal measured chemically and by flame absorption spectroscopy (Sibley et al., 1964). The inset portrays the Pb absorption at 272 nm before irradiation (dashed line) and after gamma irradiation (solid line). Although this is a powerful method, it too has limitations. First, an absorption band may not exist for the impurity in question; and, second, if the impurities aggregate into clusters or precipitates, so that the effective oscillator strength of the transition changes, then the accuracy of the method is lost.

Alkaline earth oxide crystals are much more difficult to grow in a pure state than the halides since they have much higher melting points. The most common method for producing single crystals of these materials is that of arc fusion (Butler et al., 1971; Abraham et al., 1971). The growth apparatus consists of three graphite electrodes submerged into the starting material. An arc is struck between the electrodes and this is used to melt the starting material rapidly. After several hours the current is reduced and the melt allowed to cool rather slowly. When the furnace has cooled, there is a large

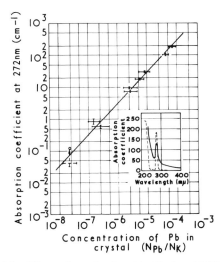

Fig. 23. The use of the 272-nm absorption band in Pb-doped KCl as a measure of lead concentration. The insert diagram shows the 272-nm band before (dashed line) and after (solid line) irradiation. After Sibley et al. (1964).

shell of sintered material in the center of the furnace which is cracked away to expose a cavity and numerous single crystals of the material. Large crystals several inches in size have been prepared in this manner. From this description of the growth technique, it should be intuitively clear that problems exist. Rapid cooling of the melted material gives rise to mechanical defects (dislocations) and the high temperatures together with the vaporizing graphite electrodes suggest chemical contamination. It has been possible to overcome these difficulties in part, however, and Table IV gives the impurity concentrations for some MgO ingots (Butler *et al.*, 1971).

TABLE IV

IMPURITY CONCENTRATIONS (μgm/gm) IN
TWO "PURE" MgO CRYSTALS[a]

Impurity	Crystal 1	Crystal 2
Al	42	41
As	< 0.3	< 0.4
Ba	< 0.7	< 0.6
Ca	47	61
Cr	< 3	< 5
Fe	3	3
Mn	0.2	0.3
N	9.2	9.3
Na	0.3	0.3
Ni	< 5	< 5
P	2	1
Pb	< 0.5	< 0.5
S	< 2	< 2
Si	27	19
Ti	3	3
Zn	7	6
Zr	< 3	< 6

[a] Grown at Oak Ridge National Laboratory.

As the need for more exotic crystals for devices has increased the art of crystal growth has evolved into a science. It is now extremely important in both basic and applied research that samples of various materials containing *known* amounts of *specific* impurities be available. This has resulted in a marked increase in the importance of crystal growth and the field has begun to receive its rightful acclaim in technology. Aside from the references already given, several excellent reviews on crystal growth are available (Gilman, 1963; Peiser, 1967).

B. Radiation Damage Processes

In this section the mechanisms by which defects can be produced in ionic solids by radiation will be discussed. It is expedient to classify the various radiation damage processes and three broad classes will be considered: electronic processes, elastic collisions, and radiolysis or photochemical mechanisms. The *electronic* class includes all processes in which no ionic defects are formed but where an electronic state is changed or charge is moved about by the absorption of radiant energy. *Elastic collisions* are those in which atoms or ions are displaced due to momentum and energy transfer from irradiating particles or photons. *Photochemical* or radiolysis processes are those in which atomic or ionic defects are formed by a series of reactions beginning with an electronic excitation. Before we consider the radiation damage processes, however, it is necessary to discuss the ways in which high energy radiation interacts with solids in the first place.

1. Interaction of Radiation with Solids

High energy photons and high energy particles interact with matter in quite different ways. Photons with energies in the range obtainable with X ray or radioisotope sources transfer their energy almost entirely to the electronic system of a crystal and in three major processes.

At low energies (0.01–0.5 MeV) it is most likely that the photon will be completely absorbed, the energy being converted into ionization and kinetic energy of one of the atomic electrons through the *photoelectric effect*. The probability of an absorption event of this kind is highest for high Z (atomic number) materials and low energy photons. The cross section varies approximately with Z^5. At energies above 0.2 MeV in all materials the electrons most likely to be excited are the K electrons of the constituent atoms. As the photon energy is reduced, the cross section for the photoelectric absorption increases, until the photon energy is too small for excitation of K shell electrons. At this point, and at lower energies when the L and M shell electrons can no longer be ionized, there is a sudden drop in cross section so that the graph of absorption coefficient versus energy has a sawtooth character in the keV energy region.

The *Compton effect*, in which a photon transfers only a portion of its energy to a crystal electron is an important energy transfer mechanism for photon energies above about 0.1 MeV. For the energy range around 1–10 MeV, the incident photon energy is about evenly distributed between the scattered photon and the electron. The total cross section for Compton scattering falls much more slowly with energy than does that of the photoelectric effect and varies only in proportion to Z, so that for the lighter elements the Compton effect is the dominant energy transfer mechanism between 0.5 and 5 MeV.

At energies above 1.02 MeV *pair production* becomes important and electron–positron pairs are produced with increasing probability as the energy rises. For energies above 10 MeV, nuclear processes can be initiated. In radiation damage research, however, it is seldom important to consider photons with energies greater than about 2 MeV since most of the isotope sources now in use are below 2 MeV in energy, e.g., [60]Co 1.17 MeV and [137]Cs 0.667 MeV.

The rate of energy transfer to a thin section of crystal by photons of a given energy is proportional to the decrease in the intensity of the irradiating beam so that

$$-dF/dx = \alpha F = F_0 \alpha \exp(-\alpha x) \qquad (14)$$

where F is the energy intensity or flux of the beam, F_0 is the initial energy flux, α is the absorption coefficient, and x is the distance from the crystal surface. Since absorption coefficients for X rays can easily reach 10–100 cm^{-1}, the variation in energy absorbed by the crystal can be as much as a factor of 10 in a few millimeters. Moreover, as noted in our previous discussion, the absorption coefficient is a strong function of photon energy. Thus, for a nonmonochromatic X-ray beam the energy absorbed is the sum of many contributions of the form of the above equation with different values of α. In this case, it is tedious to predict the pattern of energy absorption within the crystal. In the simpler case of monoenergetic photons, it is easy to calculate the pattern of primary energy absorption within a sample by a consideration of the total absorption coefficient (photoelectric + Compton + pair production) versus photon energy. The total absorption coefficient for three different metals at different photon energies is shown in Fig. 24. This figure shows that absorption coefficients

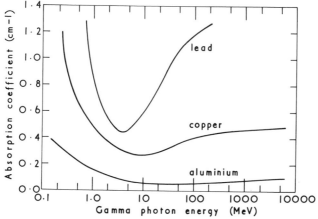

Fig. 24. Total absorption coefficient of gamma photons as a function of photon energy for Al, Cu, and Pb (from AIP handbook).

for light materials are less than for heavy materials at all energies and that a minimum occurs at an intermediate energy, as photoelectric absorption falls off and before pair production takes over. Irradiation with gamma rays having energies near the minimum clearly gives the best uniformity, and since radiation damage processes can depend nonlinearly on intensity a uniform irradiation is always useful.

In fact gamma photons give much more uniform energy deposition than softer X rays for two reasons. For many halides and oxides, as for copper in Fig. 24, the absorption coefficient for photons in the 1-MeV range is about two orders of magnitude lower than for say 100-KeV X rays, and this leads to a more uniform primary energy deposition. Less obvious is the fact that although the incident photon flux does decrease exponentially with the appropriate absorption coefficient, the scattered photons from Compton scattering cause additional energy to be transferred to the crystal as the initial beam decreases in intensity. Thus, high energy photons initiate a type of electron–photon cascade that leads to relatively uniform energy deposition throughout crystals a millimeter or more thick.

High energy *charged* particles are also used to damage crystals; so let us consider what happens when they pass through matter. The interaction is primarily coulombic with the crystal *electrons* and most of the energy loss from the particle occurs in these collisions. The rate of this energy loss can be calculated for both heavy particles and electrons and is discussed in detail in nuclear physics textbooks (Segre, 1965). From such calculations the energy loss per unit path of a heavy particle at high energies can be written as

$$-\frac{dE}{dx} = \frac{4\pi z^2 e^4 N_0 Z}{mv^2} \log \left[\frac{2mv^2}{\phi(1 - \beta^2)} - \beta^2 \right], \tag{15}$$

and for the particular case of an incident relativistic electron it is modified to read

$$-\frac{dE}{dx} = \frac{4\pi e^4}{mc^2} N_0 Z \left[\log \frac{2mc^2}{\phi} - \frac{3}{2} \log (1 - \beta^2)^{1/2} - \frac{1}{2} \log 8 + \frac{1}{16} \right] \tag{16}$$

In these equations e and m are the *electron* charge and rest mass, z and v are the charge and velocity of the incident particle, N_0 is the atomic density of the crystal, Z is the number of "excitable" electrons per target atom, and ϕ is the average excitation potential of these electrons. The ratio v/c is labeled β, c being the velocity of light.

When a heavy particle is used and its velocity decreases until it is comparable to the velocity of the crystal electrons, there is a rapid falloff in the probability of ionization events. This is partly because the maximum energy transferred to electrons is no longer sufficient to cause their ionization, that is, the number of " excitable " electrons falls dramatically, and partly because the heavy particle captures electrons from the crystal, neutralizing itself. The energy loss mechanism which then dominates is that of elastic, billiard-ball-like, collisions between the particle and the *atoms* in the crystal. Even for light particles at high energies a small fraction of the energy loss is due to elastic collisions of a similar type, but between the particle and the nuclei of the crystal atoms. This loss mechanism is never important in determining charged particle ranges, etc., but it is often important in radiation damage and will be discussed later in Section III,B,3.

The range of a particle in a crystal is extremely important in radiation damage, in deciding what is the best kind of radiation to use for a particular solid and in evaluating both the uniformity of radiation damage and the range of heavy ions for ion implantation research. Since the range depends inversely on the rate of energy loss, the above equations allow a comparison of the penetration of various heavy particles and electrons. The actual ranges of electrons and protons in aluminum for energies in the range 0.1–100 MeV is shown in Fig. 25. Ranges in other materials, with different nucleus charge Z and mass M, can be deduced from those in Al by using the dependence of dE/dx on $N_0 Z$. Thus the *mass* range R_s in a sample is related to the mass range in Al by

$$R_s = R_{Al} Z_{Al} M_s / Z_s M_{Al} \tag{17}$$

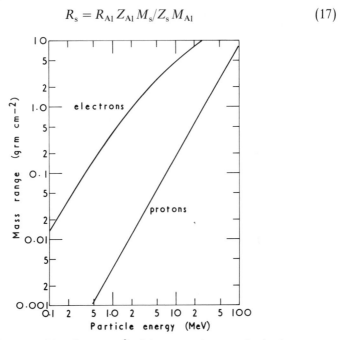

Fig. 25. The ranges (given in gm cm^{-2}) of electrons and protons in aluminum.

The range of high energy heavy particles other than protons can be estimated approximately by noting the dependence of dE/dx on z^2/v^2 and remembering that $\frac{1}{2}M_p V^2 = E_p$. Thus for a given particle energy the range of a particle p is related to that of a proton (R_{pr}) by

$$R_p = R_{pr}/Z_p^2 M_p \qquad (18)$$

Neutrons are yet another case. Since they are not charged, no electronic excitation of the crystal occurs and they interact only with nuclei via nuclear forces. On the other hand, when a *fast* neutron strikes a lattice ion, the ion can be displaced with considerable energy, acting as a charged particle moving through the lattice. This does create electronic excitation, and further elastic collisions with other crystal atoms do occur. In some materials such as LiF even *thermal* neutrons can be quite effective in producing radiation damage. When a thermal neutron is captured by ^{6}Li, charged particles are emitted which share nearly 6 MeV, and can cause damage by electronic and elastic collision processes.

2. ELECTRONIC RADIATION DAMAGE

When electrons are excited in an insulating or semiconducting crystal, the immediate result is that electron–hole pairs are produced, with the expenditure of only about 30 eV per pair. The electrons and holes are separated and move in the conduction and valance bands until they recombine or are trapped by charged defects. Since the crystal as a whole must remain neutral, for every charged defect such as an impurity there must be a compensating defect of opposite charge. For example, in MgO polyvalent impurities such as Fe, Al, Si, or Cr are charge-compensated by positive-ion vacancies. When this material is irradiated with gamma rays, mobile electron–hole pairs are produced by the radiation; these are trapped at the defect sites. The electrons might move to a Cr^{3+} ion and change it into Cr^{2+}. The holes can be trapped in the vicinity of positive-ion vacancies forming V^- centers (Hughes and Henderson, 1972). Thus, charge balance is preserved.

When the temperature of the crystal is raised, a particular set of electronic defects may become unstable and mobile. Again if we consider MgO the V^- centers are stable at 300 K but unstable at 400 K. The hole is released from the positive-ion vacancy and is retrapped at Cr^{2+}, yielding luminescence and restoring the crystal to its original state. This is a rather oversimplified example as we shall see in our detailed discussion of radiation damage in oxides, but the idea is that trapped electrons or holes on defects can dissociate. Then they will either recombine, restoring the perfect crystal; or they can be retrapped, forming different sets of electronic defects. The electronic

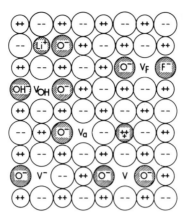

Fig. 26. Atomic models for a number of V centers in alkaline earth oxides. After Hughes and Henderson (1972).

defects formed by irradiation are color centers and their formation constitutes electronic radiation damage. Some of the various charged defects in MgO are illustrated in Fig. 26 (Hughes and Henderson, 1972).

3. ELASTIC COLLISIONS

When energetic particles such as electrons, protons, or neutrons pass through a crystal they occasionally undergo essentially elastic collisions with the nuclei of lattice atoms or ions. Simple calculations (Corbett, 1966; Seitz and Koehler, 1955; Kinchin and Pease, 1955) can be made which show that the maximum amount of kinetic energy T_m, which can be transferred from the incident particle to a lattice ion depends on the energy E and the mass M_1 of the incident particle, and the mass M_2 of the lattice ion, in the following way:

$$T_m \simeq \frac{4M_1 M_2 E(1 + E/2M_1 c^2)}{(M_1 + M_2)^2} \qquad (19)$$

In fact only for electrons is the relativistic term $E/2M_1 c^2$ significantly different from zero. Since they are particularly important, it is useful to rewrite the equation for T_m for the case of electron bombardment. If we substitute numerical values for the mass of the electron and the velocity of light then the maximum energy transfer is

$$T_m = 2147.7E(E + 1.022)/A \quad \text{(eV)} \qquad (20)$$

where E is in MeV and A is the mass number of the lattice ion.

It is clear from the general equation for T_m that heavy particles can transfer much more energy in elastic collisions than can light ones; and that electrons transfer very little energy. This does not, however, tell the whole story. Although T_m is the same for neutrons and protons (M_1 is the same in both cases), the interaction involved with the atomic nucleus is very different for the two particles; it is largely coulombic and has a very long range for protons and is nuclear, with a very short range, for neutrons. As a result of this, collisions between neutrons and atomic nuclei are much rarer than for protons, but when they do occur they are much more likely to transfer energies near T_m, so that the average transferred energy is much higher. The probability of transferring an energy E in the collision, in fact, varies with $1/E^2$ for coulombic (Rutherford) collisions so that small energy transfers are much more likely than large ones. For neutron collisions, the probability is independent of E.

There are several different ways in which atoms can be displayed by elastic collision processes, but it is clear that in any case a certain minimum amount of energy must be transferred from the incident particle to a lattice ion for displacement to occur. The amount of energy necessary if the "knock-on" lattice atom passes through the saddle point between its nearest neighbors (Seitz and Koehler, 1955), may be quite different from that required for a focusing sequence along a close packed row of like atoms (Silsbee, 1957) or for displacement through a relatively empty channel in the crystal (Holmes, 1964; Datz et al., 1967; Nelson, 1968). Nevertheless it is useful to assume that a definite amount of energy is necessary for displacement, and the formation of an interstitial–vacancy pair. This is called the displacement energy and written T_d. If the maximum energy transferred in a collision is less than the displacement threshold ($T_m < T_d$), then no elastic-collision radiation damage will occur, so that incident particles with less than a critical energy will cause no displacements and the damage rate will be strongly dependent on energy.

From the above equations for T_m, it is clear that heavier incident particles will be much more effective in displacing lattice ions than lighter ones. If sufficient energy is transferred in the primary collision, defect cascades can be produced because the displaced atom in its turn transfers sufficient energy to other lattice ions for their displacement and so on. Because the average transferred energy is high for neutron irradiation, cascades are particularly likely in this case. In cascades large clusters of defects are formed; these can change the properties of the material in a markedly different way from small defects. Detailed descriptions of elastic collision damage are available in the literature and the reader is referred to these reviews for further information (Kinchin and Pease, 1955; Seitz and Koehler, 1955; Corbett, 1966; Nelson, 1968).

The number of defects N produced by a given radiation fluence ϕ is usually written

$$N = N_0 \,\phi\sigma_D(E) \tag{21}$$

where N is the atomic density, and σ_D is the displacement cross section which is very much a function of the type and energy of the incident particles. If the sample being irradiated is thin enough, then both ϕ and σ_D will be essentially constant throughout the crystal, but for thick samples nonuniform damage can occur since one or both will change with penetration. When it is possible to measure the concentration of defects produced by the radiation and to know the flux of incident particles, then the cross section can be determined and compared with theoretical predictions. In ionic materials this concentration can often be determined by electron paramagnetic resonance, magnetic susceptibility, or a combination of magnetic and optical measurements, so that these materials are very useful in damage studies.

4. PHOTOCHEMICAL PROCESSES

For researchers who have worked mostly on metals or semiconductors, one of the surprising things about alkali halides and alkaline earth halide crystals is that radiation damage is produced in such copious quantities. Whereas in metals one might expect to produce one interstitial–vacancy pair per 2.0 MeV electron (Corbett, 1966), the same electron striking KCl can produce as many as 2000 such pairs (Schulman and Compton, 1962). In fact, it is even possible to produce Frenkel pairs by excitation with ultraviolet light (Smakula, 1930; Hall et al., 1964b; Lushchik et al., 1966). The question is, why is defect creation in highly ionic materials so efficient? It might be guessed that such a highly efficient process occurs because the energy lost by the radiation in causing electronic excitation is somehow capable of creating lattice defects by a photochemical or radiolysis process (Crawford, 1968). In such a process at least three stages must be identifiable. First, an electronic excitation must occur, resulting, at least momentarily, in creation of an electronic defect in the lattice. Second, the energy of the electronic defect must be converted into kinetic energy of a lattice ion in such a way that the ion is ejected from its normal site. Third, a route must exist for this lattice ion to move sufficiently far from its associated vacancy that a stable defect is formed. It is not easy to develop a more detailed general description of photochemical damage processes; for example, in some respects the process in MgF_2 is quite different from that in alkali halides. We will therefore leave elaborating the three stages until later.

Having described radiation damage processes in rather general terms it is now appropriate to look in more detail at what happens when real insulators are irradiated. For this, classification of the materials into three groups, alkali halides, alkaline earth halides, and oxides, is useful.

C. Alkali Halide Crystals

In alkali halides photochemical damage is extremely efficient; and since it is not possible to distinguish between photochemically produced and collision-produced Frenkel pairs, it is difficult to study radiation damage by means of elastic collisions in these materials. For example, if one were interested in surveying the effects of neutrons on KCl and decided to use a reactor as the neutron source, one would find that the gamma rays in the reactor would create far more Frenkel pairs than would the neutrons. As a result very few studies have been carried out solely on elastic collision damage in alkali halides. Therefore, our attention will be confined to electronic processes and photochemical damage. Even these two processes would be difficult to study if it were not for the fact that in insulators Frenkel pairs can be investigated optically and magnetically. If the defects were not color centers, radiation damage in ionic crystals would still be little understood.

1. ELECTRONIC PROCESSES

It has been mentioned that when a charged particle or high energy photon interacts with a crystal, lattice excitation of the crystal valence electrons is the process that occurs. Eventually the energy is redistributed in such a way that only the valence electrons are excited, and either bound or separated electron–hole pairs are formed. If the electron–hole pairs are bound, return to the ground electronic state can occur by emission of a photon or through interaction with lattice ions. When an electron is *ionized* into the conduction band and no longer bound to the hole, two processes can occur. There can be recombination of electron–hole pairs, possibly through bound excited states, or the electron can be trapped at some crystal defect. An example of the latter case is given in Fig. 27 for Tl impurity in KCl. In this instance, an electron is removed from a Cl^- ion and migrates to the Tl^+ impurity ion which has been substituted for a K^+ ion. As the electron is trapped at the Tl^+ ion to produce Tl^0, the hole that was left behind in the process is also trapped between two Cl^- ions to form a $[Cl_2^-]^+$ molecule which, if the temperature is below 175 K, will be stable in the lattice (Delbecq et al., 1961). This latter defect is produced in all alkali halides and is designated as noted in Section II as an $[X_2^-]^+$ center. In the past it has been known as the V_K center in honor of Kanzig (Kanzig, 1955; Castner et al., 1958) who first

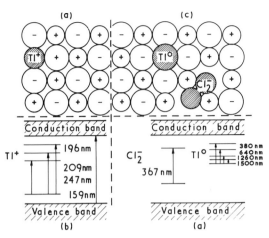

Fig. 27. Atomic structure and electron energy levels for unirradiated and irradiated KCl: Tl; (a) atomic model of Tl^+ in KCl; (b) energy levels and optical transitions for Tl^+ in KCl; (c) atomic models of Tl^0 and $[Cl_2^-]^+$ in KCl; (d) energy levels and optical transitions for (c). Wavelength values are taken from Delbecq *et al.* (1966).

studied it. There are numerous examples of charge trapping in the alkali halides and, besides Tl (Delbecq *et al.*, 1966, 1967) Ag, Au, Mn, and many other impurities have been characterized (Gebhardt and Mohler, 1966a,b; Gomes, 1963; Fischer, 1969; Itoh and Ikeya, 1970; Ikeya, 1972). In general, it is possible to study these impurities by means of optical absorption and emission measurements and various valence states can usually be identified. Some defects such as Mn are paramagnetic and it is then possible to use electron paramagnetic resonance to complement optical methods in investigating the impurity (Itoh and Ikeya, 1970).

Eppler (1961) has catalogued the emission bands from various impurities in alkali halide crystals, but detailed characterization involving absorption, emission, and magnetic resonance have rarely been done. This is partly because the photochemical damage mechanism is so efficient in these materials that it becomes difficult to distinguish between ionic defects and electronic defects. Nevertheless, from the studies that have been completed one very important observation has been made. In the alkali halides electrons can move freely once they have been ionized into the conduction band, but holes can move only with a diffusive motion, hopping from one lattice site at which they are trapped at an $[X_2^-]^+$ center to the next, and at low temperatures they do not move at all (Delbecq *et al.*, 1961; Sonder and Sibley, 1972). The motion of these holes in various lattices has been studied extensively in the last few years and it appears that the activation energy for motion of the holes depends on the strength of the halogen–halogen bond and the relative sizes of the anion and cation in the lattice. The activation energy for LiF is about 0.32 eV (Kanzig, 1960), for KCl about 0.54 eV (Keller *et al.*, 1967), for NaI about 0.15 eV (Murray and Keller, 1967) and for KI for about 0.27 eV (Keller and Murray, 1966). As we will see the $[X_2^-]^+$ center or self-trapped hole also plays a major role in the photochemical damage process and the temperature at which it becomes mobile, i.e., its activation energy, is crucial (Sonder and Sibley, 1972).

2. PHOTOCHEMICAL PROCESSES

a. Intrinsic Coloration. It has been known for many years that X rays can very efficiently produce coloration in alkali halide crystals (Pohl, 1937). Initially, it was not known if this coloration was due to electronic defects such as valence changes of impurities or to ionic defects such as vacancies and interstitials. However, the postulate of the correct structure of the F center (de Boer, 1937) and its identification as the defect responsible for the coloration of these materials indicated that more than electronic effects were involved.

Discussions of the color center production mechanism then centered on whether interstitials must also be produced or whether the F centers were formed at dislocation lines (Seitz, 1954). The definitive evidence of the EPR work of Kanzig and Woodruff (1958) showed that for *low temperature irradiation* interstitials were indeed formed; as H centers which are $[X_4^{3-}]$ molecular ions located symmetrically on three lattice sites along $\langle 110 \rangle$ lines of anions. This research was soon supported by that of Compton and Klick (1958) who demonstrated optically that H centers were produced at low temperatures in proportion to the number of F centers formed and concluded that F centers and H centers were produced simultaneously by radiation. These results were convincing enough that soon there was general agreement that at low temperatures, Frenkel pairs were produced as intrinsic defects. It was not until considerably later, however, that a combination of flow stress and optical measurements by Nadeau (1962, 1963, 1964) and others (Sibley and Sonder, 1963; Sibley and Russell, 1965) and especially the lattice expansion measurements of Balzer *et al.* (1968a,b) provided convincing evidence that Frenkel pair production by X rays occurred at all radiation temperatures. Only very recently (Hobbs *et al.*, 1972a,b, 1973) has more detailed knowledge of the fate of the interstitial atoms produced at room temperature become available.

Despite this evidence that intrinsic Frenkel pair production occurred at all radiation temperatures, the production mechanism itself was still not clear; none of the three stages we outlined in Section III,B,4 had been defined. Several observations led three independent groups of investigators to arrive at the conclusion that the production mechanism was intimately associated with the *nonradiative* recombination of electrons with $[X_2^-]^+$ centers (Crawford, 1968). The groups were led by Hersh in the U.S., Lushchik in the U.S.S.R. and Pooley in the U.K. The observations that stimulated the theories were the symmetry of the $[X_2^-]^+$ center (Kabler, 1964; Murray and Keller, 1965), and the fact that intrinsic coloration sometimes occurs even when ultraviolet light is used as the incident radiation (Hall *et al.*, 1964b; Lushchik *et al.*, 1965), and the spectacular dependence of F center production in KI on temperature (Fig. 28) (Hall *et al.*, 1964a). Evidence for connecting F-center production with electron recombination with $[X_2^-]^+$ centers accumulated when Pooley (1966b) showed that for several alkali halides the F-center production efficiency was dependent on whether electron–hole recombination luminescence occurred or not. When emission occurred, F-center production decreased.

Definitive evidence that electron–hole recombination does produce F centers and H centers has now been obtained by Keller and Patten (1969), and

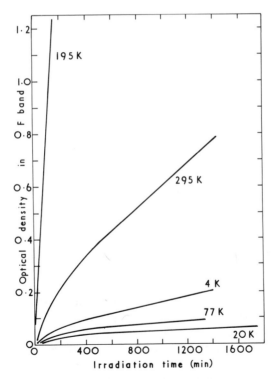

Fig. 28. F band growth in KI under X irradiation at five different temperatures. After Hall *et al.* (1964a).

was described in Section II,D,2. In the meantime Lushchik and his co-workers (1965) and Goldstein (1967) had shown that F centers could be produced by ultraviolet light with enough energy only to excite bound electron–hole pairs and insufficient for ionization. No wonder this process is so efficient!

It may be useful to pause at this point to emphasize the important role color centers have played in determining the intricasies of the damage mechanism in alkali halides. As we have seen and shall see as we progress further, the damage process in these materials is rather complex. If it had not been for the multitude of tools available to study defects in these insulators and the fact that many defects give rise to optical and magnetic transitions, very little progress would have been made. It is possible to define the particular type of defect under study, e.g. vacancy, vacancy-pair, interstitial, trapped hole, etc., and even to determine its orientation in the crystal lattice (as noted in Section II,C,6) by means of optical or magnetic measurements. The advances that have been made serve to illustrate again how color centers are powerful tools with which to investigate the microscopic details of radiation damage.

Having established that intrinsic coloration occurs at all temperatures, that the $[X_2^-]^+$ center electron pairs are the electronic defects that are precursors to F center production, and that nonradiative recombination of electrons with $[X_2^-]^+$ centers provided the mechanism for converting electronic energy into kinetic energy, it remained to show how sufficiently large F center–H center separations could be achieved to prevent immediate recombination. How do the H centers and F centers separate? Pooley (1966a) postulated that, because of the $\langle 110 \rangle$ orientation of the $[X_2^-]^+$ center in the lattice and the close proximity of the two halide ions forming the center, the nonradiative recombination transition causes the two halide ions to be thrown apart along the $\langle 110 \rangle$ directions, roughly sharing the recombination energy. An anion replacement collision sequence would then be propagated along the close-packed halide row. It is now clear (Ueta et al., 1969) that the ejection process is slightly more complex because the halide ion is actually ionized in leaving the vacancy (Pooley and Hatcher, 1970; Smoluchowski et al., 1971), but it still seems likely that a replacement sequence provides the means for F center and H center separation.

Two major pieces of evidence for replacement sequences in F center production can be cited. If no such sequences occurred, then a sample irradiated at low temperature (4–30°K) with X rays would be expected to contain only fairly close pair defects since the interstitials are not mobile and the nonradiative recombination events cannot impart sufficient energy to the lattice ions to drive them directly very far away from their original lattice sites. It would follow that, when the temperature was raised to where the interstitials become mobile, they would be likely to recombine with their vacancy. If this happened, two observations should be evident. First, very few, if any, residual F centers would remain after the annealing; and second, when samples were irradiated at the annealing temperature, little, if any, radiation damage would occur. In alkali halide crystals neither effect is observed, which suggests that the interstitials are too far from the vacancies to recombine in a correlated fashion. In MgF_2 both effects occur, and replacement sequences are in any case unlikely (Buckton and Pooley, 1972).

If focusing replacement collisions do occur along $\langle 110 \rangle$ directions, then the substitution of larger or smaller anions or cations in the crystals should disrupt these collisions and reduce the coloration rate. Data taken on mixed alkali halide crystals do indeed show this effect (Fig. 29) (Still and Pooley,

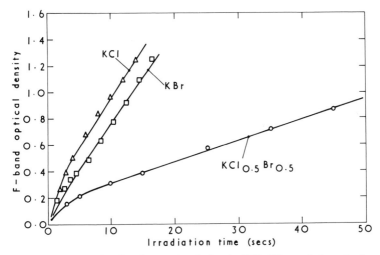

Fig. 29. F center creation by electron irradiation in KCl, KBr, and the mixed crystal $KCl_{0.5}Br_{0.5}$. Irradiations carried out at 400 KeV, 0.4 μA cm^{-2} at ~ 10 K.

1969; Hirai, 1972), providing further evidence that collision sequences are important in the coloration process in alkali halides. This is not to say that no coloration can occur without these collision sequences, only that they seem to increase the coloration efficiency of alkali halides.

When irradiations are made at liquid helium temperature all defects are immobile after creation and only the collision sequence discussed above allows the vacancies and interstitials to be separated appreciably. However, at temperatures nearer 300 K, first H centers, then $[X_2^-]^+$ centers and finally F^+ centers become mobile. As each of these defects become mobile the radiation damage process is affected (Sonder and Sibley, 1972). In most halides the interstitial atom, the H center, becomes mobile first. As it does so, it can recombine with vacancies or be trapped by impurities and other interstitials. Hobbs and co-workers (1972a,b, 1973) have clear electron microscope evidence for some of these interstitial clusters in alkali halides. Similar events occur when vacancies become mobile, which is always at higher temperatures than for interstitials. They aggregate to form F_2 and F_3 centers and they can annihilate interstitials or be trapped by impurities. Since interstitials cluster even at low temperatures (Hobbs *et al.*, 1972a,b, 1973), when vacancies do eventually become mobile only other vacancies and impurities are widely distributed in the lattice. This means that moving vacancies are much more likely to interact with impurities (to form F_A or Z like centers) and with other vacancies (to form F aggregate centers) than with interstitials. It should be pointed out at this stage that the common mobile vacancy entity is the F^+ center (Lüty, 1968; Link and Lüty, 1965; Hartel and Lüty, 1964a,b). Thus, when a crystal is irradiated at room temperature and left in the dark, no aggregation of F centers occurs, even after months. However, if the sample is optically bleached with light in the F band, vacancies do move and F center aggregation occurs. This is because the bleaching process momentarily forms F^+ centers from F centers and these are the defects that move.

Some of the most dramatic effects on the radiation damage process in alkali halides arise from the behavior of holes and of nonradiative electron hole recombination, because the hole (or $[X_2^-]^+$ center) is the precursor of the photochemical damage process. Thus at low temperatures coloration is often suppressed because radiative recombination predominates over nonradiative recombination (Pooley, 1966a). At high temperatures the $[X_2^-]^+$ centers become mobile, and are able to move through the lattice and recombine with electrons at impurities rather than in the undisturbed lattice. When this happens, the production process is inhibited (Pooley, 1966b, 1968).

An illustration of the effects of interstitial trapping, hole mobility and vacancy mobility, we will consider how lead in KCl affects the coloration. Figure 30 shows the coloration at 80 K of KCl: Pb samples doped with various amounts of Pb (Sonder and Sibley, 1965a). Note that the F center concentration is greater in the more heavily doped samples. This is because the impurities trap interstitials before they can recombine with vacancies. On the other hand, when the radiation temperature is high enough that $[X_2^-]^+$ centers are mobile, Pb suppresses the coloration (Sibley *et al.*, 1964) (Fig.31). In this instance, many of the mobile holes recombine with electrons at impurities rather than in the undisturbed lattice; thus reducing the number of Frenkel pairs produced. Yet another effect arises because the Pb impurities act as electron traps, so that during the radiation process the ratio of F^+ centers to F centers is larger than in a pure crystal. The higher

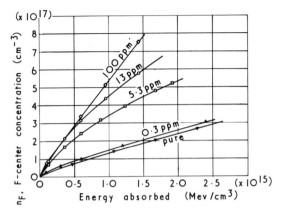

Fig. 30. The formation, by irradiation with 1.5-MeV electrons, of F centers in lead doped KCl at ∼80 K. After Sonder and Sibley (1965a).

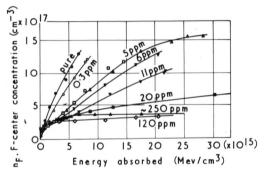

Fig. 31. F band growth caused by 1.5-MeV electron irradiation of lead-doped KCl at room temperature. After Sibley *et al.* (1964).

mobility of the F^+ center causes enhanced recombination with the impurity trapped interstitials. It might be expected that an increase in radiation intensity could overcome this suppression and this is indeed the case (Pooley, 1966b; Sibley *et al.*, 1964).

b. Extrinsic Coloration. In addition to those intrinsic coloration processes we have been discussing, other *extrinsic* mechanisms exist, whereby color centers can be produced in the vicinity of crystal imperfections by radiation (Crawford, 1968). This type of coloration is perhaps the most studied and least understood of all radiation damage processes. It has been shown that there are probably a number of specific mechanisms active in the extrinsic process (Alvares-Rivas and Levy, 1967; Sanchez and Agullo-Lopez,

1968). The coloration depends on the type and concentration of the impurities and on deformation (Sonder and Sibley, 1972). In the case of deformation it appears that dislocation debris and the formation of vacancy clusters by dislocation–dislocation intersections and cross slipping play a more important role in the coloration process than the dislocations themselves. Much more needs to be done in this exceedingly complex research area, but it is clear that any further experiments must be well thought out and pursued with great care (Agullo-Lopez and Jaque, 1973).

D. Alkaline Earth Halide Crystals

1. ELECTRONIC PROCESSES

Since these materials have potential as both phosphors and laser hosts (Nassau, 1971), a number of studies have been made that deal with electronic processes in them. Also, CaF_2 has been thoroughly investigated as a possible photochromic material (Faughnan *et al.*, 1971). The situation is qualitatively like that in alkali halides in that, as well as the normal valence changes that occur when impurities are excited by light or high energy photons, it is possible to form $[X_2^-]^+$ centers in CaF_2, and $KMgF_3$. Hayes (1970) and his collaborators have also studied self-trapped holes in SrF_2 and BaF_2. As in alkali halides, a large concentration of impurities that trap electrons is useful in allowing the production of numerous $[X_2^-]^+$ centers. Some of the photochromic processes will be discussed in Section IV.

2. PHOTOCHEMICAL DAMAGE

Just as in alkali halides and alkaline earth oxides, optical and magnetic measurements of F centers in materials such as MgF_2, $KMgF_3$, and CaF_2 provide a convenient and useful way to monitor the radiation damage (Hayes, 1970). The damage mechanism in many alkaline earth halides is almost as efficient as it is in alkali halides and shows no dependence on the energy of the incident particle. This indicates that photochemical damage dominates in these materials and that it would again be difficult to investigate damage produced by elastic collisions.

Crystals with the rutile structure such as MgF_2 are significantly different from alkali halides in that no close packed rows of anions exist along which focused collision sequences can be propagated. This means that collision sequences cannot be longer than one or two lattice dimensions. This is useful information because radiation damage studies in these materials and in alkali halides allow a comparison of cases where focused collisions can and cannot occur. If no focused collision sequences are allowed, then it is expected that radiation events will create close pairs, and as the temperature is increased, annihilation of correlated vacancy–interstitial pairs will occur. This is what happens in irradiated MgF_2. Figure 32 shows the annealing stages for MgF_2 crystals irradiated at 7 K (Sibley and Facey, 1968). Notice that there are annealing stages at about 70 K, where 70% of the defects anneal out, and at 150 K. The explanation of these stages is the following (Sibley, 1971; Buckton and Pooley, 1972). In a low temperature irradiation interstitial–vacancy pairs are created by the radiation with some of the interstitials being trapped in near-neighbor sites and some being forced farther into the lattice and ending up several lattice spaces away. As the temperature is increased, the near-neighbor interstitials recombine at about 70 K and since most of the damage is of this type approximately 70%

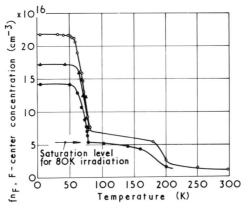

Fig. 32. Annealing of the F band in three samples of MgF_2 irradiated at ~ 7 K. The samples were warmed at 25 deg K/min below 80 K. Annealing above 80 K was carried out by heating quickly to the required temperature, maintaining that temperature for 10 min and recooling to 80 K for measurement. After Sibley and Facey (1968).

of the damage is annealed out. The interstitials that are produced at greater distances still feel the effect of the strain field around the vacancy since the potential energy barrier surrounding the interstitial is reduced in the direction of the vacancy. The rate at which the interstitial jumps is

$$v_J = v_0 \exp\left(-E_J/kT\right) \tag{22}$$

where v_j is the number of jumps occurring per second, v_0 is the "attempt" frequency ($\sim 5 \times 10^{12}$ sec^{-1} in MgF_2), k is Boltzman's constant, and E_J is the jump activation energy. If the activation energy for interstitial motion is reduced by about 0.06 eV in the direction of a vacancy compared to what it would be for a jump away from the vacancy, then at 100 K it will be 10^3

times more likely to jump toward the vacancy than away from it. On the other hand, at 250 K an interstitial will be only 30 times more likely to jump in the direction of the vacancy and may escape to be trapped by another interstitial or an impurity. From this analysis, we might expect that if irradiation is carried out as a function of temperature very little stable damage might occur for irradiations below 200 K, but stable damage could be formed above this temperature. This is exactly what is observed as is shown in Fig. 33 (Sibley and Facey, 1968). These results give us confidence that the

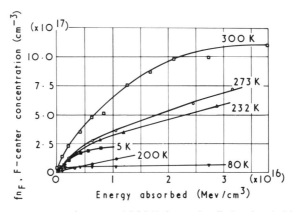

Fig. 33. F center concentration versus 1.7-MeV electron irradiation does in MgF$_2$ crystals at six different temperatures. After Sibley and Facey (1968).

above model is correct. More refined experiments by Buckton and Pooley (1972) further substantiate this model. By assuming the above model and working out the kinetics for the process, they were able to show that the predictions of the model fit their more detailed measurements of F center production in the range between 50 and 200 K.

Radiation damage in KMgF$_3$ samples has also been studied. These crystals have the perovskite structure and focused sequences can occur, although not as effectively as in alkali halides. Figure 34 portrays the F center production in this material as a function of radiation temperature (Riley and Sibley, 1970), and shows that the temperature dependence of the damage process is similar to that in KCl. More work must be done on this material to increase our understanding of the importance of focused collision sequences.

In spite of the work of Hayes and his collaborators (Hayes, 1970) on CaF$_2$ and the numerous studies of F centers in this material, it is not yet possible to state that unambiguous evidence for the production of F centers in CaF$_2$ by radiation has been obtained (Görlich et al., 1968; Hayes, 1970). It is certainly possible to produce $[X_2^-]^+$ centers in this material, and it might be expected that F centers should be formed. In undoped SrF$_2$ and BaF$_2$ both

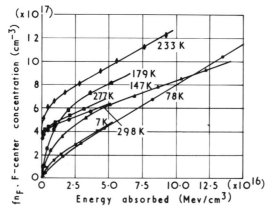

Fig.34. F center concentration versus electron irradiation does in KMgF$_3$ at seven different temperatures. After Riley and Sibley (1970).

$[X_2^-]^+$ centers and F centers have been observed in crystals irradiated with X rays at 77 K, but these centers anneal out when the samples are warmed to room temperature. The radiation damage behavior of SrF$_2$ down to ~ 5 K has been examined in detail by Hayes and Lambourn (1973), who find that F centers, interstitial ions, and $[X_2^-]^+$ centers are formed, rather than F and H centers as in alkali halides. It is possible, however, that processes similar to those occurring in MgF$_2$ are also important in the fluorite structure and that correlated recombination of defects occurs. For example, Ratnam (1966) believes he has evidence that F centers can be produced in CaF$_2$ at high radiation temperatures. This would suggest a similarity to MgF$_2$ damage with the annealing stages moved to higher temperature.

E. Alkaline Earth Oxide Crystals

These materials have the same crystal structure (interpenetrating face-centered cubic) as the alkali halides, but they are divalent and their ionicity is less. For these reasons, it is interesting to compare radiation damage in MgO with that in the alkali halides. It should be mentioned that, in addition to being extremely useful in unraveling the radiation damage processes in materials, color centers are also valuable in studying electron–lattice interactions and crystal fields. This latter aspect of color center work in oxides has been excellently covered by Hughes and Henderson (1972). In radiation damage studies in oxides no evidence for a photochemical process of coloration has been reported; as we said at the beginning of Section III, it is very unlikely that photochemical damage could occur. Therefore, in our coverage of this subject we will emphasize electronic processes and Frenkel pair production by elastic collisions.

1. ELECTRONIC PROCESSES

Recently grown oxide crystals are much purer than those grown several years ago. Nonetheless even the purer ones contain appreciable amounts of impurity and, because charge compensation must occur, intrinsic lattice defects are also always present. For example, two substitutional Cr^{3+} ions can be compensated by one positive-ion vacancy. This type of compensation, which must occur in all crystals, gives rise to the possibility of many different charged defects when the materials are irradiated. The movement and trapping of charge at these defects is strictly an electronic process since no ionic defects are created or annihilated, and the material can be returned to its original state by thermal decomposition of the trapped centers.

Perhaps these effects are best illustrated by considering what happens in MgO upon gamma irradiation and subsequent thermal annealing. This is shown in Fig. 35 for four different impurities and the V^- center (a positive-ion vacancy with one trapped hole) (Sibley et al., 1969). As shown by the dashed portions of the curves, room temperature gamma irradiation produces V^- centers, causes the divalent vanadium concentration to increase, and reduces the concentration of the three other ions. Heat treatment above 800 K almost completely restores the preirradiation charge distribution. As these changes in valence state occur, there are also marked changes in the optical properties of the crystals. At about 400 K a red thermoluminescence emission occurs (Wertz et al., 1967) that is most likely due to the process

$$Cr^{2+} + e^+ \rightarrow Cr^{3+*} \rightarrow Cr^{3+} + hv_1 \qquad (23)$$

where e^+ represents a hole and hv_1 is the emitted photon.

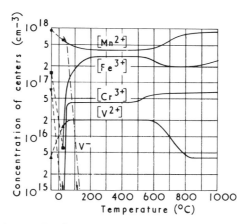

Fig. 35. Effects of gamma irradiation and thermal annealing on impurity bands and the V^- center in MgO. The dashed curves on the left show the effects of irradiation, the subsequent full curves, and the dot–dash curve for V^-, show the effect of thermal annealing. After Sibley et al. (1969).

A correlation of the luminescence emission at a particular temperature with a particular defect impurity reaction is difficult since a very small number of emitting atoms can give a strong luminescence. Sophisticated measurements are necessary; for example, in thermoluminescence measurements it has become standard practice to measure not only simple glow curves of total light intensity versus sample temperature, but also the spectra of the light emitted in each glow peak. It is also helpful if the charge states of many impurities can be changed by means other than radiation. A reducing atmosphere such as hydrogen can change the valence of Fe^{3+} to Fe^{2+} when the sample is at high temperature, and conversely a high temperature oxidation treatment can reverse the process. It is therefore extremely useful to compare the emission from heat-treated and irradiated specimens. Many impurity spectra have been previously identified, and Johnson (1966) has reviewed the effects of valence changes on the luminescence of several oxygen-dominated lattices.

V^- centers are useful in studying elastic collision damage in oxides since they give a measure of the number of positive-ion vacancies present in the lattice. However, they have been particularly useful in studying charge transfer in these materials. The presence of V^- centers in irradiated MgO was first reported by Wertz *et al.* (1959). Later Wertz and his co-workers (1967) found that thermal destruction of V^- centers at 400 K leads to the red Cr^{3+} luminescence mentioned above.

2. ELASTIC COLLISIONS

When MgO crystals are irradiated with neutrons or high energy electrons an absorption band at 250 nm which is due to both F^+ and F centers appears (Sibley and Chen, 1967; Sonder and Sibley, 1972; Hughes and Henderson, 1972). When the same samples are gamma irradiated, no F centers are formed. On the other hand, additively colored samples heat treated in magnesium vapor at 2000 K, contain only F centers and not interstitials. Figure 36 is a plot of the normalized absorption coefficient for electron irradiated, neutron irradiated, and additively colored specimens (Chen *et al.*, 1968), and it shows the subtle shape changes due to changing the relative concentration of F^+ and F centers. If photochemical processes were active in MgO, it might be expected that gamma rays would also produce Frenkel pairs, but in causing elastic collision displacements they are relatively ineffective.

The identification of the absorption bands of simple defects greatly simplifies detailed studies of the radiation damage processes. For example, monitoring the 250-nm band in MgO as a function of irradiation intensity, type, and energy can give a great deal of information about the damage process in MgO. Thus the optical studies of the energy dependence of the

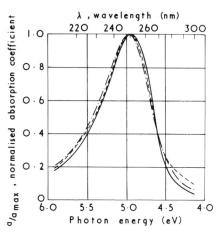

Fig. 36. The shapes of the 250-nm band in MgO containing F and F^+ centers, measured at room temperature: neutron irradiated, $\alpha_{max} = 250$ cm^{-1}; electron irradiated, $\alpha_{max} = 25$ cm^{-1}; additively colored, $\alpha_{max} = 370$ cm^{-1}. After Chen *et al.* (1968).

cross section (Chen *et al.*, 1970) show that the threshold for F center production corresponds to irradiation with ~ 0.33-MeV electrons. From this observation and the equation for T_m given in Section III,B,3, a displacement threshold energy of about 60 eV for the oxygen ion can be computed. This is close to the 50 eV found for Al_2O_3 samples (Compton and Arnold, 1961), and seems reasonable. However, a displacement threshold of 60 eV leads theoretically to a cross section for displacement of oxygen by 1.7 MeV electrons of 11.6 b. The experimental cross section, again measured using F^+ and F bands to give defect concentration, is only 1.7 b. The difference between these values suggests that the net F center production rate at room temperature depends not simply on the elastic collision processes but also on the recombination and stabilization processes that occur simultaneously. This is confirmed by two other observations. First, the fact that few radiation-induced V^- centers are observed unless the radiation is performed at low temperature even though the theoretical cross section of magnesium is about the same as for oxygen. Second, the observation that Fe-doped MgO damages more easily than does pure material (Sibley and Chen, 1967). It is therefore likely that oxygen interstitials are mobile at room temperature and can recombine with vacancies unless they are trapped by impurities or other defects. The greater the impurity concentration, the more stabilized interstitials and the larger the observed F center concentration. The marked energy dependence of the radiation damage rate is evidence for elastic collision processes being the dominant damage mechanism since photochemical processes are essentially independent of the energy of the incident particle.

Not only is it possible to use color centers to gain information on how defects are produced by radiation, but they can also be used to study recombination processes. An example of this is illustrated by the data shown in Fig.37 (Chen *et al.*, 1969). From these data it is clear that defects in irradiated specimens anneal out at lower temperatures than in additively colored samples. This suggests that in irradiated crystals the interstitials trapped during the radiation process in clusters or at impurities are released around 600 K and recombine with vacancies. On the other hand, the fact that the F centers in additively colored material do not anneal out until 1200 K indicates that the vacancies are not mobile until this temperature.

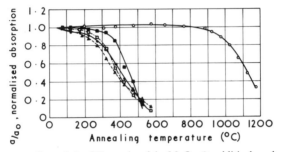

Fig. 37. The annealing of the 250-nm band in MgO; ○ additively colored crystals; □ electron irradiated crystals; ▣, ●, △ crystals neutron irradiated under different conditions. After Chen *et al.* (1969).

Fewer radiation damage experiments have been performed on other alkaline earth oxides, but indications are that elastic collisions are the dominant damage mechanism. Neutron irradiation or irradiation with very high energy electrons produces many more cluster centers than do 2.0-MeV electrons, but the damage kinetics are very similar.

IV. Applications of Color Centers

The level of financial support for scientific research has increased enormously over the past 25 years. Not surprisingly, as society has paid the piper more handsomely it has become more determined to call the tune, at least to the extent of insisting that most of the scientific research now carried out should have worthwhile objectives in mind. More and more, scientists are asked to explain the relevance of their work. It is therefore pertinent for us now to ask what is, or can be, the value of the science of color centers and radiation damage to our society. In this section we provide a partial answer by discussing how color center science contacts important technological areas.

A. Radiation Dosimetry

High energy radiation is now widely used in medical therapy and radiography, in industrial radiography, for sterilization of food and medical instruments, and in a thousand other ways. Moreover, it is almost certain that we will have to rely on nuclear fission and fusion, with the accompanying high energy radiation, as the prime sources of energy as we move into the twenty-first century. This means that an increasing number of people will be in possible contact with radiation and their exposure to it must be simply, cheaply, and carefully monitored. Solid state integrating dosimeters, in which changes in the dosimeter material caused by radiation can be measured easily and at sufficiently low doses (in practice the minimum dose which needs to be measured is about 100 mrad $= 10^{-6}$ J/gm^{-1}) are the best answer to this need. Three methods are widely used at present. One is based on the blackening of photographic film by the radiation, which is a complex color center process. The other two methods, radiation-induced photoluminescence and thermoluminescence, use color centers more directly.

Defect photoluminescence was discussed in Section II; it provides one very sensitive way of detecting defects. Its sensitivity can be illustrated by a comparison with defect absorption. An absorbed dose of 100 mrad $(6 \times 10^{12}$ eV gm$^{-1})$ in a sample of reasonable size would create at most $\sim 10^{11}$ defects by an electronic or photochemical process. If the created defects had a strong absorption band like the F center, the sample would absorb only $\sim 10^{-2}\%$ of the incident light. Therefore, in detecting these centers by absorption, we would need measurements of light intensity to $10^{-2}\%$ or better which is very difficult. On the other hand, the same sample with $\sim 10^{11}$ defects when excited with light of intensity of about 1 mW would give a luminescence signal of $\sim 2 \times 10^{11}$ photons sec^{-1} (assuming a luminescence quantum efficiency of 1) and measurement of such a signal is trivially easy. A further feature of most *defect photoluminescence* which is a big advantage for dosimetry is that the excitation wavelength is usually significantly smaller than that for emission. This allows the luminescence and the scattered exciting light to be separated with a monochromator or a filter, so that the ratio of signal to background can be high.

The use of radiation-induced photoluminescence was pioneered by Schulman *et al.* (1951) for U.S. Navy purposes. The materials that they suggested and which are still the most widely used are silver-doped phosphate glasses, although many improvements have now been made (Yokota and Nakajima, 1965) which allow dose measurements down to 10 mrad. A typical dosimeter glass is made from a base containing 47% LiPO$_3$ and 53% Al(PO$_3$)$_3$ to which 12% AgPO$_3$ is added. Irradiation creates color centers that have an absorption band in the near uv at about 320 nm (Fig. 38) and excitation at

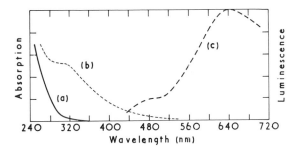

Fig. 38. Absorption and luminescence of silver activated phosphate glass: (a) absorption of unirradiated glass; (b) absorption of irradiated glass; (c) luminescence of irradiated glass caused by excitation at 320 nm. After Schulman (1967).

320nm causes efficient luminescence at 640 nm. The emission intensity is a well-defined and fairly linear function of dose (Fig. 39) and the intensity does not change on standing at room temperature after the first few hours. Many millions of these dosimeters have been used.

The mechanism of operation of these dosimeters has been deduced from a color center study of silver-activated alkali halides (Etzel and Schulman, 1954). Although for a variety of reasons these materials are less suitable for dosimetry, they behave in the same way as do the glasses and the structure of the important defects is much easier to determine in the single alkali halide crystals than in structureless glasses. In the radiation process single substitutional Ag^+ ions act as hole traps and become Ag^{2+}. Neighboring pairs of Ag^+ ions, which may have trapped anion vacancies, act as electron traps and

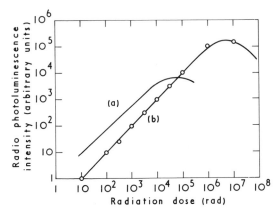

Fig. 39. The intensity of photoluminescence in (a) silver activated phosate glass and (b) "pure" LiF as a function of radiation dose. Glass data are from Schulman (1967), LiF data from Regulla (1972). Note that the absolute values of intensity for the two materials are not necessarily related as indicated on the diagram.

form centers of type shown in Fig. 40. These latter centers absorb at 310 nm in KCl and emit at 556 nm; they are analogous to the photoluminescence centers used in the glasses. These dosimeters are therefore useful examples of the purely electronic effects of irradiation of insulators discussed in Section III.

Photoluminescence dosimetry is also possible using pure alkali halides (Schulman *et al.*, 1951; Regulla, 1972). The recent results obtained in LiF by Regulla are shown in Fig. 39. It is not clear that the LiF photoluminescence system would be satisfactory under all conditions (it is not used at present)

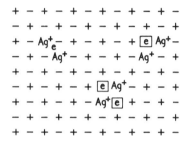

Fig. 40. Possible atomic structures of the photoluminescence (C) centers in Ag^+ doped KCl.

since it is only accidentally linear. The F_2 centers are responsible for the observed emission and these are created by irradiation in proportion to the square of the number of primary defects, the F centers (Faraday *et al.*, 1961; Sonder and Sibley, 1965b). In LiF, however, the F center concentration grows with the square root of the dose (Durand *et al.*, 1969) so that linear photoluminescence results.

In radiation-induced thermoluminescence the stored energy of the defects is used for their detection. When the irradiated dosimeter material is heated, the defects, which may be ionic or simply electronic, recombine and luminescence occurs. The sensitivity of this system is potentially higher than for photoluminescence since no exciting light is necessary (eliminating the need for filters or monochromators) and photon counting can be used. Probably the most widely used thermoluminescent material is LiF containing magnesium. Daniels (Boyd and Daniels, 1949) pioneered the use of this material and it is desirable because of its close approximation in X ray stopping power to human tissue. The range of doses that can be measured using LiF or $Li_2B_2O_7$ thermoluminescence is truly remarkable (Fig. 41).

The mechanism of thermoluminescence in LiF is still not very clear. Magnesium impurities are essential for proper behavior, and it is very likely that the different thermoluminescence peaks correspond to different aggregations of Mg^{2+} and positive-ion vacancies, the aggregates acting as hole traps (Claffy, 1965). Klick *et al.*, (1967) proposed that heating caused these traps to

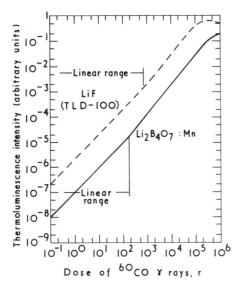

Fig. 41. Thermoluminescence output as a function of radiation dose for LiF and $Li_2B_4O_7$. After Schulman (1967).

release the holes, which then recombined with electrons in F centers to give luminescence. However, although the F center may act as a recombination center in LiF containing only Mg as a deliberate impurity it is less effective than the centers formed by small additions of a second impurity such as Ti (Zimmerman and Jones, 1967; Rossiter *et al.*, 1970, 1971). In spite of considerable research effort in this area the thermoluminescence mechanism in LiF : Mg : Ti is far from being properly determined, and a rather more complicated proposal has been made by Mayhugh (1970; see also Christy and Mayhugh, 1972).

Nevertheless, the increasingly widespread use of thermoluminescence in LiF for dosimetry (Attix, 1972) is another good example of the usefulness of color center generation in alkali halides which probably involves the photochemical damage mechanism discussed in Section III.

B. Display Devices

Since the father of all color centers is the alkali halide F center, it is appropriate that this center was the first to find an application, in the Skiatron display tube. The Skiatron tube was developed during the Second World War (Rosenthal, 1940; Kazan and Knoll, 1968). It consisted of a fairly ordinary cathode ray tube (crt) but with a KCl layer in place of the phosphor. The writing electron beam created F centers in the KCl, and the absorption of visible light by these centers produced a "dark trace" which could easily be seen from the front. The F centers were bleached steadily in the incident viewing light, but only slowly compared with the decay in brightness of a normal crt. The simple KCl Skiatron was not, in the end, very successful, but "dark trace" tubes do have many attractive properties which have prompted the examination of other materials for the purpose. For example, the contrast of a "dark trace" picture does not change with incident light intensity; it is as easy to see in bright sunlight as in a normally lit room. It is, therefore, much better than a light-emitting crt for use in aircraft, ships, or even in well-lit rooms. Another big advantage is that the color centers composing the picture can be arranged to decay as slowly as required. This can be particularly important for computer driven displays, where the information in the display is usually changed fairly slowly, because the effective memory of the display tube itself makes it unnecessary to initiate the expensive process of having the computer refresh the display more often than every second or so. Normal crts require refreshing every 20–50 msec. This feature is also important for radar, where the information input to the display usually comes from a slowly rotating mechanical aerial.

Some of the recent dark trace display tubes use F centers created in bromide sodalite ($Na_6Al_6Si_6O_{24} \cdot 2NaBr$) (Phillips and Kiss, 1968; Taylor et al., 1970). The production of F centers in sodalites is phenomenologically very like that in alkali halides, although mechanisms that are different in detail must be involved. F centers are definitely produced in two very different stages (Fig. 42) analogous to the fast and slow stages of coloration of alkali halides (Faughnan and Shidlovsky, 1972). As in alkali halides the

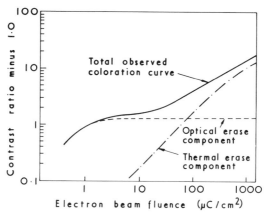

Fig. 42. The production of F centers in photosensitive sodalite by 20-kV electron irradiation. The measurement is of diffuse reflectance contrast ratio which gives a good indication of F center concentration. After Faughnan and Shidlovsky (1972).

fast stage is largely the result of added impurities and in it F centers are produced with very high efficiencies ($\sim$ 100 eV/F). Indeed, early efforts to use sodalite in dark trace displays concentrated on the fast stage F centers, with the advantages of high writing speed and easy optical erasure in mind. These failed largely because of the interference of slow stage F centers which could not be erased optically.

As a result workers at RCA now favor the use of fairly "pure" sodalite, where the fast stage coloration is suppressed and slow stage F centers can be produced more efficiently. Even in this mode sodalite is much better than KCl in that a greater density of F centers is possible, giving a much better optical contrast between the dark trace and white background. Slow stage F center production is also slightly more efficient in sodalite than in alkali halides and, more important for display applications, does not begin to saturate until very high defect concentrations are reached. It can also be made fairly linear with electron fluence (Fig. 43) (Faughnan and Shidlovsky, 1972). Although the images written on sodalite tubes have to be removed by heating (Heyman *et al.*, 1971) this can be achieved in only a few seconds. Thermal erasure by electron beam heating is also a possibility (Hankins *et al.*, 1972), and would give the display tube a selective erase facility.

C. *Electron Beam Memories*

The computer designer would like to be able to use high capacity memories ($>$ 100 Mbits) which have short access time ($<$ 10 μsec), high data input and output rates ($>$ 5 MHz) and low cost ($<$ 10^{-2} ¢/bit). In practice a

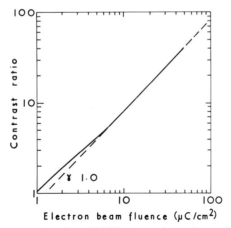

Fig. 43. Contrast ratio as a function of 25-kV electron beam fluence for bromide sodalite, showing the excellent linearity achievable with substrate heating. The full line is the experimental curve and the dashed line that for $\gamma = 1.0$. After Faughnan and Shidlovsky (1972).

choice must be made between memories with the right speeds but rather high cost (~ 0.5 ¢/bit), such as semiconductor or magnetic core memory, and others which have the right costs but much greater access times (~ 50 msec) such as magnetic drums or disks. Electron beam accessed memories using color centers could provide one approach to a more ideal memory, and one concept of this kind of device is being examined at Harwell.

The phenomenology of the Harwell memory plane (Bishop *et al.*, 1972) is as follows. The plane consists of a layer (which may be single crystal, powder, or evaporated film) of a halide based phosphor such as KI : Tl. Access to it is by a 10–30-kV electron beam that can be directed to any one of 10^6 or more locations on the plane. Efficient photochemical radiation damage can occur when the electron beam is operating at medium current, creating defects which reduce the initially high efficiency of the phosphor by a factor of up to 100–1000. Since it is possible to measure the efficiency of the phosphor by exciting it with the same electron beam, but at a much lower current, a pattern of information can easily be written on and retrieved from the memory plane. Furthermore, if the beam is used at very high current the defects introduced at medium current can be destroyed, with consequent restoration of the luminescence. The behavior of the Harwell plane is summarized in Fig. 44.

Simply because it is an electron beam memory the Harwell system has a potentially short access time. The 50-msec time typical of magnetic disk arises because access involves first mechanically moving a magnetic pick-up head across the surface of a disk and then waiting for the appropriate part of the disk to pass under the head. In an electron beam system, on the other hand, access simply involves positioning the beam which can be done in a few hours. The major advantage that the Harwell system has over other reversible electron beam accessed memories is that the plane is initially homogeneous and therefore potentially cheap to produce. Apart from cheapness in manufacture of the plane itself the fact that the electron beam determines it own array of locations means that requirements of electron, optical, and mechanical precision in the tube are also considerably relaxed, with consequent cost savings. Another obvious advantage is that any of the 10^6–10^8 locations is coupled optically to a single output channel for read-out, with consequent ease of processing. The electrical isolation provided by the optical coupling is also an important feature, and it seems likely that the cost per bit for a final system will lie between 10^{-4} and 10^{-2} ¢.

The Harwell memory is like the sodalite display in that it relies on the creation and destruction of F centers and complementary interstitial atoms for its operation, but it differs in actually using the interstitials rather than the F centers. When interstitials atoms are created in KI : Tl, they are very

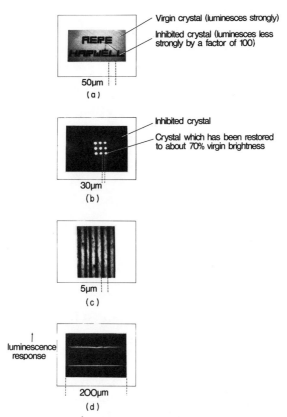

Fig. 44. The phenomenology of the Harwell computer memory plane. (a) Electron beam inhibition, time taken to inhibit ~ 10 μsec; (b) electron beam restoration, time taken to restore luminescence ~ 10 μsec; (c) storage density, locations can be as little as 5 μm apart–equivalent to a storage density of 4 Mbit/cm²; (d) material uniformity KI(Tl) shows a response that is uniform to better than 10%.

efficiently trapped by thallium ions to give thallium–halogen atom complexes. Although these complexes remain efficient recombination centers for electron–hole pairs, and of course it is electron–hole recombination at thallium ions which causes the luminescence of the virgin crystal, they do not emit luminescence in the process. The complexes also act as traps for further interstitial atoms and the result is that the luminescence efficiency of the irradiated crystal falls exponentially with dose (Fig. 45).

The destruction of the thallium interstitial complexes cause the return of efficient luminescence. Although this can be achieved thermally, the times required are very long; many milliseconds even near the melting point of KI : Tl and several seconds at ~ 400°C. In fact it is possible to destroy the

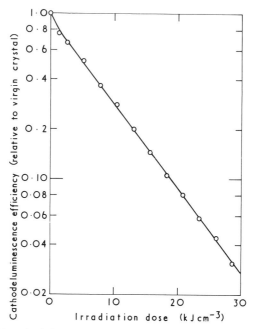

Fig. 45. The fall in cathode-luminescence efficiency of KI : Tl with electron irradiation dose.

complexes by electron beam heating much more quickly; luminescence restoration times of a few microseconds are easily achieved although no part of the crystal exceeds 400°C. This occurs (Bishop *et al.*, 1972) because of radiation-enhanced processes. In this case it is likely that the stability of the thallium–interstitial complexes is greatly reduced when they are in an excited electronic state.

The Harwell memory plane provides a good example of how photochemical color center creation can be used for information storage in computer memory devices.

D. *Optical Memories*

Rapid access to any part of a memory plane can also be achieved using a light beam. In practice the only source of light of sufficient intensity for data storage is the laser, and electro-optical or accousto-optical methods must be used for sufficiently rapid deflection of the beam. Since lasers are bulky and expensive and deflectors are unlikely to be able to access more than 10^5–10^6 locations, direct bit-by-bit optical information storage is not at present of great interest for high speed systems. The cost can be reduced only by moving the storage medium under the light beam, but large access times are then unavoidable. However, holographic data storage can be used. With holography each location accessed by the light deflector can be a hologram, which can be used to reconstitute and array (say 10^4–10^5) data bits as shown in Fig. 46. The result is a system that can in principle store $10^{10}+$ bits at very low cost and with rapid access. Holographic data storage also has the advantage that it provides redundancy in the use of storage material, and this can be very useful in eliminating reading errors. Thus the area required to record a given picture holographically is about $2\frac{1}{2}$ times that required to record it directly with the same signal to noise ratio (Ramberg, 1972). The overall promise of optical memories is great (Rajchman, 1972) and a large number of optical memory materials are available. All of these have some specific areas of advantage but few are as easily reversible or have the high resolution of the color center examples. The following discussion is limited

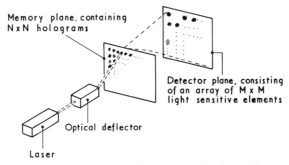

Fig. 46. A schematic diagram of readout from an optical holographic memory. The laser beam is deflected on to any one of N^2 holograms, each hologram giving a reconstructed image of M^2 bits on the detector plane.

to two of the possible color centers candidates: F_A type centers in CaF_2 and M centers in alkali halides.

The F_A center in alkali halides is an F center which has one of the nearest neighbor alkali ions replaced by a different alkali ion, for example by Na^+ in KCl. In CaF_2 containing La or Ce, and which has been additively colored in Ca vapor, F centers associated with rare earth impurities are formed (Staebler and Schnatterly, 1971). Using the F_A center notation these centers might be called F_{RE}, with RE representing the nearest neighbor rare earth ion. The F_{RE} centers have broad absorption bands in the violet and near uv region by which they can be ionized. Since the crystals also contain trivalent rare earth ions, uv irradiation in the F_{RE} bands causes the following transformation to occur (Staebler and Kiss, 1969):

$$F_{RE} + RE^{3+} \xrightarrow{h\nu} F_{RE}^+ + RE^{2+}$$

The process can be reversed by photon irradiation in the RE^{2+} absorption band, which is usually in the visible region, and the state of the crystal can, in principle at least, be sensed using lower energy transitions of the F_{RE} center which are in the red and near ir region of the spectrum and do not cause ionization of the center. The absorption spectrum of additively colored CaF_2 : La before and after uv irradiation at 200–400 nm is shown in Fig. 47,

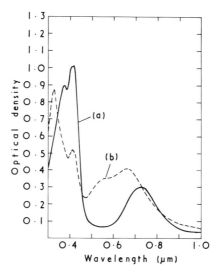

Fig. 47. The optical absorption spectra of photochromic CaF_2 : La : Na: (a) Stable state; (b) after irradiation in the range 380–460. After Duncan (1972).

on which the write, erase, and read bands can be seen (see Duncan, 1972, for other examples).

The quantum efficiency of the forward photochromic process, that is the ionization of F_{RE}, is thought to approach 1, and that of the reverse process is comparable, so that the sensitivity is very good. The principle disadvantages of this system are that a high concentration of centers can never be achieved, so that thick samples have to be used to give reasonable optical densities, and that the metastable state does not have enough stability to retain information for a long time. For example, cerium doping produces centers that are stable for about 1 week at room temperature. It seems very likely that this instability arises from the tunneling of electrons from RE^{2+} to F_{RE}^{+}.

The centers in CaF_2 are straightforward photochromic centers. In contrast, the use of the M center proposed by Bron *et al.*, (1965) and independently by Schneider (1967) is quite different. The M center in alkali halides was introduced in Section II. It consists of two neighboring F centers, has a $\langle 110 \rangle$ symmetry axis (van Doorn, 1962), and exhibits two absorption bands. One is traditionally called the M band and is excited by light having an electric vector *parallel* to the long axis of the center (Fig. 48). Centers excited by M band light either luminesce or decay nonradiatively, but in neither case is the M center changed by the process. The second optical band of the center, commonly called the M_F because it is often hidden by the F band, is at higher energy and has a transition moment *perpendicular* to the long axis of the center (Fig. 48). Light in the M_F band

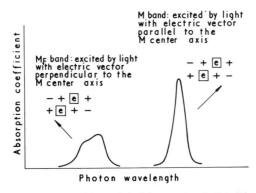

Fig. 48. The transitions of the M center in alkali halides.

causes the M centers to reorient into another of the six possible orientations.

The way in which the M center system can be used to store digital information is fairly clear. For example, a binary 1 can be represented by having all of the centers lying on the [110] direction and a binary 0 by having all the centers on [1$\bar{1}$0]. If a beam of M_F light polarized along [110] is directed onto a particular location, it will excite all the centers other than those lying along [110]. However, when any center in another orientation is excited it may reorient into the [110] position. Since no return from the [110] orientation is possible because these centers are not excited, eventually all the centers adopt this orientation. In this way a binary 1 can be written. Writing the binary 0 requires [1$\bar{1}$0] polarized light. Readout can be achieved nondestructively using polarized light in the M band; if the reading light is [110] polarized, a location in the 1 state absorbs strongly, whereas one in a 0 state does not absorb. An alternative readout technique is possible because the reorientation process is complex and involves ionization of the M center (Schneider, 1970). The extent of ionization can be reduced with infrared illumination, which prevents the electrons produced by M center ionization remaining trapped elsewhere, and the efficiency of reorientation by M_F light is also reduced. With this effect almost nondestructive readout can be achieved using M_F light, and only one laser need be used.

One problem with the M center in alkali halides is that reorientation implies motion of the center, and motion eventually leads to loss of centers or other undesirable "fatigue" effects. However, if M_A centers are used, rotation can occur around the impurity, and fatigue is avoided.

E. The Nuisance of Radiation-Induced Defects

Parts A–D of this section have concentrated on the positive usefulness or potential of color centers in devices. However, the study of color centers and other radiation induced defects does have bearing on technology in another way: radiation-induced degradation. There are two particularly important examples where radiation damage is paramount in determining the overall economics, namely nuclear reactors and communication satellites. In both cases radiation-induced defects are a great nuisance and limit operating life.

Fuel elements, cladding, and other structural materials near the core of a power reactor are subject to an intense gamma and fast neutron flux. Naturally the materials chosen are always ones that do not damage easily,

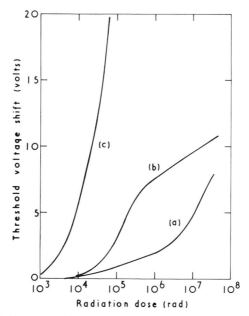

Fig. 49. Threshold voltage shifts in MOS devices under gamma irradiation. Aluminium gates were used with (a) zero bias, (b) −20 V bias, and (c) +20 V bias during irradiation. Data from Lindmayer and Noble (1968).

certainly none in which photochemical damage occurs can be used. Even so, the slow buildup of defects over the operating life of the reactor does cause slow dimensional changes, mechanical failure, and corrosion; these effects must be minimized for efficient operation over many years. For this reason the behavior of all kinds of point defects in possible reactor materials and also in simpler "model" materials have been studied, particularly in metals and in ceramic oxides such as MgO. Ceramic materials in particular offer the possibility of high temperature operation, with accompanying increases in thermal efficiency of the reactor. Most advanced reactor fuels are oxides or carbides; their behavior under irradiation is of paramount importance.

Communications satellites and, for that matter, earth resource satellites and other spacecraft, always carry complex electronic equipment. This equipment is bombarded continuously in space by high energy electrons and protons emitted by the sun. The radiation produces ionization events in the oxide layers of MOS (metal–oxide–semiconductor) devices and the electrical charges are trapped by defects in the oxide. The result is that the electrical characteristics of the MOS transitors, particularly the voltage at which the devices can be turned on or off, change with time (Lindmayer and Noble, 1968) (Fig. 49). This is another area where color centers are a nuisance and knowledge of their behavior important.

References

Abragam, A., and Bleaney, B. (1970). "Electron Paramagnetic Resonance of Transition Ions." Oxford Univ. Press (Clarendon), London and New York.

Abraham, M. M., Butler, C. T., and Chen, Y. (1971). *J. Chem. Phys.* **38**, 2166.

Agullo-Lopez, F., and Jaque, F. (1973). *J. Phys. Chem. Solids* **34**, 1949.

Alvares-Rivas, J. L., and Levy, P. W. (1967). *Phys. Rev.* **162**, 816.

Attix, F. H. (1972). *Health Phys.* **22**, 287.

Balzer, R., Peisl, H., and Waidelich, W. (1966). *Phys. Status Solidi* **15**, 495.

Balzer, R., Peisl, H., and Waidelich, W. (1968a). *Phys. Status Solidi* **28**, 207.

Balzer, R., Peisl, H., and Waidelich, W. (1968b). *Phys. Lett. A* **27**, 31.

Bessent, R. G. (1969). *Proc. Phys. Soc., London (Solid State Phys.)* **2**, 1101.

Binder, D., and Sturm, W. J. (1954). *Phys. Rev.* **96**, 1519.

Bishop, H. E., Henderson, R. P., Irdale, P., and Pooley, D. (1972). *Appl. Phys. Lett.* **20**, 504.

Bowen, D. H., and Clarke, F. J. P. (1964). *Phil. Mag.* **9**, 413.

Boyd, C. A., and Daniels, F. (1949). *J. Chem. Phys.* **17**, 1221.

Bron, W. E., Dreyfus, R. W., and Heller, W. R. (1965). U.S. Pat. No. 3,466,616.

Bucci, C., and Fieschi, R. (1964). *Phys. Rev. Lett.* **12**, 16.

Buckton, M. R., and Pooley, D. (1972). *Proc. Phys. Soc., London (Solid State Phys.)* **5**, 1553.

Butler, C. T., Russell, J. R., Quincy, R. B., and LaValle, D. E. (1966). *J. Chem. Phys.* **45**, 968.

Butler, C. T., Sturm, B. J., and Quincy, R. B. (1971). *J. Cryst. Growth* **8**, 197.

Castner, T. G., and Kanzig, W. (1957). *Phys. Chem. Solids* **3**, 178.

Castner, T. G., Kanzig, W., and Woodruff, T. O. (1958). *Nuovo Cimento, Suppl.* **7**, 612.

Chandra, S., and Rolfe, J. (1970). *Can. J. Phys.* **48**, 412.

Chen, Y., Sibley, W. A., Srygley, F. D., Weekes, R. A., Hensley, E. B., and Kroes, R. L. (1968). *J. Phys. Chem. Solids* **29**, 863.

Chen, Y., Williams, R. T., and Sibley, W. A. (1969). *Phys. Rev.* **182**, 960.

Chen, Y., Trueblood, D. L., Schow, O. E., and Tohver, H. T. (1970). *Proc. Phys. Soc., London (Solid State Phys.)* **3**, 2501.

Christy, R. W., and Mayhugh, M. R. (1972). *J. Appl. Phys.* **43**, 3216.

Claffy, E. W. (1965). *Proc. Int. Conf. Lumin. Dosim., Stanford, Calif., U.S. At. Energy Comm.* **CONF-650637**, 74.

Clarke, C. D., Mitchell, E. W. J., and Stewart, J. H. (1971). *Cryst. Lattice Defects* **2**, 105.

Compton, W. D., and Arnold, G. W. (1961). *Discuss. Faraday Soc.* **31**, 130.

Compton, W. D., and Klick, C. C. (1958). *Phys. Rev.* **110**, 349.

Compton, W. D., and Rabin, H. (1964). *Solid State Phys.* **16**, 121.

Cooke, J. S., and Dryden, J. S. (1962). *Proc. Phys. Soc., London* **80**, 479.

Corbett, J. W. (1966). *Solid State Phys., Suppl.* **7**.

Cottrell, A. H. (1953). "Dislocations and Plastic Flow in Crystals." Oxford Univ. Press (Clarendon), London and New York.

Crawford, J. H. (1968). *Advan. Phys.* **17**, 93.

Datz, S., Erginsoy, C., Leibfried, G., and Lutz, M. O. (1967). *Amer. Rev. Nucl. Sci.* **17**, 129.

Dawson, R. K., and Pooley, D. (1969). *Phys. Status Solidi (B)* **35**, 95.

de Boer, J. H. (1937). *Rec. Trav. Chim. Pays-Bas* **56**, 301.

Delbecq, C. J., Hayes, W., and Yuster, P. H. (1961). *Phys. Rev.* **121**, 1043.

Delbecq, C. J., Ghosh, A. K., and Yuster, P. H. (1966). *Phys. Rev.* **151**, 599.

Delbecq, C. J., Ghosh, A. K., and Yuster, P. H. (1967). *Phys. Rev.* **154**, 797.

Duncan, R. C. (1972). *RCA Rev.* **33**, 248.

Durand, P., Farge, Y., and Lambert, M. (1969). *J. Phys. Chem. Solids* **30**, 1353.

Elliott, R. J., and Taylor, D. W. (1964). *Proc. Phys. Soc., London* **83**, 189.

Eppler, R. A. (1961). *Chem. Rev.* **61**, 523.

Eshelby, J. D. (1953). *J. Appl. Phys.* **24**, 1249.

Eshelby, J. D. (1954). *J. Appl. Phys.* **25**, 255.

Estermann, I., Leivo, W. J., and Stern, O. (1949). *Phys. Rev.* **75**, 627.

Etzel, H. W., and Schulman, J. H. (1954). *J. Chem. Phys.* **22**, 1549.

Faraday, B. J., Rabin, H., and Compton, W. D. (1961). *Phys. Rev. Lett.* **7**, 57.

Farnum, E. H., and Royce, B. S. H. (1968). *Phys. Lett. A* **26**, 164.

Faughnan, B. W., and Shidlovsky, I. (1972). *RCA Rev.* **33**, 273.

Faughnan, B. W., Staebler, D. L., and Kiss, Z. J. (1971). *Appl. Solid State Sci.* **2**, 107.

Feher, G. (1957). *Phys. Rev.* **105**, 1122.

Feher, G. (1959). *Phys. Rev.* **114**, 1219, 1245.

Fischer, F. (1969). *Z. Phys.* **231**, 293.

Fitchen, D. B. (1968). *In* "Physics of Color Centers" (W. B. Fowler, ed.), p. 293. Academic Press, New York.

Fitchen, D. B., Fetterman, M. R., and Pierce, C. B. (1966). *Solid State Commun.* **4**, 205.

Fleischer, R. L. (1962a). *Acta Met.* **10**, 835.

Fleischer, R. L. (1962b). *J. Appl. Phys.* **33**, 3504.

Fowler, W. B. (1968). *In* "Physics of Color Centers" (W. B. Fowler, ed.), p. 54. Academic Press, New York.

Friedel, J. (1963). "Dislocations." Addison-Wesley, Reading, Massachusetts.

Fuller, R. G. (1972). *In* "Point Defects in Solids" (J. H. Crawford and L. M. Slifkin, eds), p. 103. Plenum, New York.

Gebhardt, W., and Kuhnert, M. (1964). *Phys. Lett.* **11**, 15.

Gebhardt, W., and Mohler, E. (1966a). *Phys. Status Solidi* **14**, 149.

Gebhardt, W., and Mohler, E. (1966b). *Phys. Status Solidi* **15**, 255.

Gilman, J. J., ed. (1963). "The Art and Science of Growing Crystals." Wiley, New York.
Görlich, P., Karras, H., Symanowski, C., and Ullmann, P. (1968). *Phys. Status Solidi* **25**, 93.
Goldstein, F. T. (1967). *Phys. Status Solidi* **20**, 379.
Gomes, W. (1963). *Trans. Faraday Soc.* **59**, 1648.
Grundig, H. (1965). *Z. Phys.* **182**, 477.
Guggenheim, H. (1961). *J. Appl. Phys.* **32**, 1337.
Hall, T. P. P., Pooley, D., and Wedepohl, P. T. (1964a). *Proc. Phys. Soc., London* **83**, 635.
Hall, T. P. P., Pooley, D., Runciman, W. A., and Wedepohl, P. T. (1964b). *Proc. Phys. Soc., London* **84**, 719.
Hankins, M. C. A., Procter, R., and Hughes, G. (1972). *Electron Lett.* **8**, 278.
Hartel, H., and Lüty, F. (1964a). *Z. Phys.* **177**, 369.
Hartel, H., and Luty, F. (1964b). *Z. Phys.* **182**, 111.
Hayes, W. (1970). *In* "Nonmetallic Crystals" (S. C. Jain and L. T. Chadderton, eds.), p. Gordon & Breach, New York.
Hayes, W., and Lambourn, R. F. (1973). *Proc. Phys. Soc. London (Solid State Phys.)* **6**, 11.
Henderson, B. H., and Bowen, D. M. (1971). *Proc. Phys. Soc., London (Solid State Phys.)* **4**, 1487.
Henderson, B. H., and King, R. D. (1966). *Phil. Mag.* **13**, 1149.
Henderson, B. H., and Wertz, J. E. (1968). *Advan. Phys.* **17**, 750.
Hersh, H. N. (1966). *Phys. Rev.* **148**, 928.
Heyman, P. M., Gorog, I., and Faughnan, B. (1971). *IEEE Trans. Electron Devices* **18**, 685.
Hirai, M. (1972). *Solid State Commun.* **10**, 493.
Hobbs, L. W. (1970). *J. Sci. Instrum.* **3**, 85.
Hobbs, L. W., Hughes, A. E., and Pooley, D. (1972a). *U.K. At. Energy Auth., Res. Group, Rep.* **AERE-R6953**.
Hobbs, L. W., Hughes, A. E., and Pooley, D. (1972b). *Phys. Rev. Lett.* **29**, 234.
Hobbs, L. W., Hughes, A. E., and Pooley, D. (1973). *Proc. Roy. Soc., Ser. A* **332**, 167.
Holmes, D. K. (1964). *In* "The Interaction of Radiation with Solids" (R. Strumane, J. Nihoul, R. Gevers, and S. Amelinckx, eds.), p. 147. North-Holland Publ., Amsterdam.
Huang, K., and Rhys, A. (1950). *Proc. Roy. Soc., Ser. A* **204**, 406.
Hughes, A. E. (1968). *J. Phys. Chem. Solids* **29**, 1461.
Hughes, A. E., and Henderson, B. (1972). *In* "Point Defects in Solids" (J. H. Crawford and L. M. Slifkin, eds.), p. 381. Plenum, New York.
Hughes, A. E., and Runciman, W. A. (1965). *Proc. Phys. Soc.,* **86**, 615.
Ikeya, M. (1972). *Phys. Status Solidi (B)* **51**, 407.
Itoh, N., and Ikeya, M. (1970). *In* "Nonmetallic Crystals" (S. C. Jain and L. T. Chadderton, eds.), p. 161. Gordon & Breach, New York.
Ivey, M. F. (1947). *Phys. Rev.* **72**, 341.
Jensen, P. (1939). *Ann. Phys. (Leipzig).* **34**, 161.
Johnson, P. D. (1966). *In* "Luminescence of Inorganic Solids" (P. Goldburg, ed.), p. 287. Academic Press, New York.
Johnston, W. G., and Nadeau, J. S. (1964). Air Force ARL Rep. 64-135. Contract No. AF33(616)-7942.
Kabler, M. N. (1964). *Phys. Rev. A* **136**, 1296.
Kabler, M. N. (1972). *In* "Point Defects in Solids" (J. H. Crawford and L. M. Slifkin, eds.), p. 327. Plenum, New York.
Kanzig, W. (1955). *Phys. Rev.* **99**, 1890.
Kanzig, W. (1960). *J. Phys. Chem. Solids* **17**, 80.
Kanzig, W., and Woodruff, T. O. (1958). *J. Phys. Chem. Solids* **9**, 70.
Kaplyanski, A. A. (1964). *Opt. Spektrosk.* **16**, 602, 1031. [*Opt. Spectrosc. (USSR)* **16**, 329, 557 (1964).]
Karlov, N. V., Margerie, J., and d'Aubigné, Y. M. (1963). *J. Phys. (Paris),* **24**, 717.

Kazan, B., and Knoll, M. (1968). "Electronic Image Storage." Academic Press, New York.
Keller, F. J., and Murray, R. B. (1966). *Phys. Rev.* **150**, 670.
Keller, F. J., and Patten, F. W. (1969). *Solid State Commun.* **7**, 1603.
Keller, F. J., Murray, R. B., Abraham, M. M., and Weeks, R. A. (1967). *Phys. Rev.* **154**, 812.
Kinchin, G. H., and Pease, R. S. (1955). *Rep. Progr. Phys.* **18**, 1.
Klick, C. C. (1972). *In* "Point Defects in Solids" (J. H. Crawford and L. M. slifkin, eds.),
    p. 291. Plenum, New York.
Klick, C. C., Claffy, E. W., Gorbics, S. G., Attix, F. H., Schulman, J. H., and Allard, J. G. (1967).
    J. Appl. Phys. **38**, 3867.
Lindmayer, J., and Noble, W. P. (1968). *IEEE Trans. Electron Devices* **15**, 637.
Link, E., and Lüty, F. (1965). *Int. Symp. Color Cent. Alkali Halides, Urbana, Ill.*
Lüty, F. (1968), *In* "Physics of Color Centers" (W. B. Fowler, ed.), p. 182. Academic Press,
    New York.
Lüty, F., and Mort, J. (1964). *Phys. Rev. Lett.* **12**, 45.
Lushchik, C. B., Liidya, G. G., and Elango, M. A. (1965). *Sov. Phys.—Solid State* **6**, 1789.
Lushchik, C. B., Vale, G. K., Ilmas, E. R., Rooze, N. S., Elango, A. A., and Elango, M. A. (1966).
    Opt. Spectrosc. (USSR) **21**, 377.
McGowan, W. C., and Sibley, W. A. (1969). *Phil. Mag.* **19**, 967.
Markham, J. J. (1959). *Rev. Mod. Phys.* **31**, 956.
Markham, J. J. (1966). *Solid State Phys., Suppl.* **8**.
Martin, D. G. (1962). *Appl. Mater. Res.* **1**, 160.
Martin, D. G. (1968). *Proc. Phys. Soc., London (Solid State Phys.)* **1**, 333.
Martin, D. G., and Henson, D. W. (1964). *Phil. Mag.* **9**, 659.
Mayhugh, M. R. (1970). *J. Appl. Phys.* **41**, 4776.
Möstoller, M., Ganguly, B. N., and Wood, R. F. (1971). *Phys. Rev. B* **4**, 2015.
Mollwo, E. (1931). *Nachr. Ges. Wiss. Goettingen, Math.-Phys. Kl.* p. 97.
Molner, J. P. (1941). *Phys. Rev.* **59**, 944. (Abstr. No. 157.)
Morgan, C. S., and Bowen, D. H. (1967). *Phil. Mag.* **19**, 165.
Murray, R. B., and Keller, F. J. (1965). *Phys. Rev. A* **137**, 942.
Murray, R. B., and Keller, F. J. (1967). *Phys. Rev.* **153**, 993.
Nadeau, J. S. (1962). *J. Appl. Phys.* **33**, 3480.
Nadeau, J. S. (1963). *J. Appl. Phys.* **34**, 2248.
Nadeau, J. S. (1964). *J. Appl. Phys.* **35**, 1248.
Nassau, K. (1961). *J. Appl. Phys.* **32**, 1820.
Nassau, K. (1971). *Appl. Solid State Sci.* **2**, 174.
Nelson, R. S. (1968). "The Observation of Atomic Collisions in Crystalline Solids." North-
    Holland Publ., Amsterdam.
Nowick, A. S. (1972). *In* "Point Defects in Solids" (J. H. Crawford and L. M. Slifkin, eds.),
    p. 151. Plenum, New York.
Ottmer, R. (1928). *Z. Phys.* **46**, 798.
Peech, J. M., Bower, D. A., and Pohl, R. O. (1967). *J. Appl. Phys.* **38**, 2166.
Peiser, H. S., ed. (1967). "Crystal Growth." Pergamon, New York.
Peisl, H., Balzer, R., and Waidelich, W. (1966). *Phys. Rev. Lett.* **17**, 1129.
Petroff, S. (1950). *Z. Phys.* **127**, 443.
Phillips, W., and Kiss, Z. (1968). *Proc. IEEE* **56**, 2072.
Pohl, R. W. (1937). *Proc. Phys. Soc., London* **49**, 3.
Pooley, D. (1965). *Solid State Commun.* **3**, 241.
Pooley, D. (1966a). *Proc. Phys. Soc., London* **87**, 245, 257.
Pooley, D. (1966b). *Proc. Phys. Soc., London* **89**, 723.
Pooley, D. (1968). *Proc. Phys. Soc., London (Solid State Phys.)* **1**, 323.
Pooley, D., and Hatcher, R. D. (1970). *Bull. Amer. Phys. Soc.* **15**, 340.

Pratt, P. L., Chang, R., and Newey, C. W. A. (1963). *Appl. Phys. Lett.* **3**, 83.
Rajchman, J. A. (1972). *J. Vac. Sci. Technol.* **9**, 1151.
Ramberg, E. G. (1972). *RCA Rev.* **33**, 5.
Ratnam, V. V. (1966). *Phys. Status Solidi* **16**, 559.
Regulla, D. F. (1972). *Health Phys.* **22**, 491.
Riley, C. R., and Sibley, W. A. (1970). *Phys. Rev. B* **1**, 2789.
Rosenthal, A. H. (1940). *Proc. IRE* **28**, 203.
Rossiter, M. J., Rees-Evans, D. B., and Ellis, S. C. (1970). *Brit. J. Appl. Phys.* **3**, 1816.
Rossiter, M. J., Rees-Evans, D. B., Ellis, S. C., and Griffiths, J. M. (1971). *Brit. J. Appl. Phys.* **4**, 1245.
Runciman, W. A. (1965). *Proc. Phys. Soc., London* **86**, 629.
Sanchez, C., and Agullo-Lopez, F. (1968): *Phys. Status Solidi* **29**, 217.
Schneider, I. (1967). *Appl. Opt.* **6**, 2197.
Schneider, I. (1970). *Phys. Rev. Lett.* **24**, 1296.
Schulman, J. H. (1967). *Proc. Int. Conf. Lumin. Dosim., Stanford, Calif., U.S. At. Energy Comm.* **C-650637**.
Schulman, J. H., and Compton, W. D. (1962). "Color Centers in Solids." Macmillan, New York.
Schulman, J. H., Ginter, R. J., Klick, C. C., Alger, R. S., and Levy, R. A. (1951). *J. Appl. Phys.* **22**, 1479.
Schumacher, R. (1970). "Introduction to Magnetic Resonance." Benjamin, New York.
Segre, E. (1965). "Nuclei and Particles." Benjamin, New York.
Seidel, H., and Wolf, H. C. (1968). *In* "Physics of Color Centers" (W. B. Fowler, ed.), p. 538. Academic Press, New York.
Seitz, F. (1954). *Rev. Mod. Phys.* **26**, 7.
Seitz, F., and Koehler, J. S. (1955). *Solid State Phys.* **2**, 305.
Sibley, W. A. (1971). *IEEE Trans. Nucl. Sci.* **18**, 273.
Sibley, W. A., and Chen, Y. (1967). *Phys. Rev.* **160**, 7120.
Sibley, W. A., and Facey, O. E. (1968). *Phys. Rev.* **174**, 1076.
Sibley, W. A., and Russell, J. R. (1965). *J. Appl. Phys.* **36**, 810.
Sibley, W. A., and Sonder, E. (1963). *J. Appl. Phys.* **34**, 2366.
Sibley, W. A., Sonder, E., and Butler, C. T. (1964). *Phys. Rev. A* **136**, 537.
Sibley, W. A., Kolopus, J. L., and Mallard, W. C. (1969). *Phys. Status Solidi (B)* **31**, 223.
Silsbee, R. H. (1957). *J. Appl. Phys.* **28**, 1246.
Smakula, A. (1930). *Z. Phys.* **63**, 762.
Smoluchowski, R., Lazareth, O. W., Hatcher, R. D., and Dienes, G. J. (1971). *Phys. Rev. Lett.* **27**, 1288.
Sonder, E. (1962). *Phys. Rev.* **125**, 1203.
Sonder, E., and Sibley, W. A. (1965a). *Phys. Rev. A* **140**, 539.
Sonder, E., and Sibley, W. A. (1965b). *Phys. Status Solidi* **10**, 99.
Sonder, E., and Sibley, W. A. (1972). *In* "Point Defects in Solids" (J. H. Crawford and L. M. Slifkin, eds.), p. 201. Plenum, New York.
Sonder, E., and Walton, D. (1967). *Phys. Lett. A* **25**, 222.
Spalt, H., and Peisl, H. (1971). *Int. Conf. Color Cent. Ionic Cryst., Reading, Eng.*
Staebler, D. L., and Kiss, Z. J. (1969). *Appl. Phys. Lett.* **14**, 93.
Staebler, D. L., and Schnatterly, S. E. (1971). *Phys. Rev. B* **3**, 516.
Still, P. B., and Pooley, D. (1969). *Phys. Status Solidi (B)* **32**, K147.
Stoneham, A. M. (1966). *Proc. Phys. Soc., London* **89**, 909.
Stott, J. P., and Crawford, J. H. (1971). *Int. Conf. Color Cent. Ionic Cryst., Reading, Eng.*
Taylor, M. J., Marshall, D. J., Forrester, P. A., and MacLaughlan, S. D. (1970). *Radio Electron. Eng.* **40**, 17.
Trinkhaus, H., Spalt, H., and Peisl, H. (1970). *Phys. Status Solidi (A)* **2**, K97.

Ueta, M. (1952). *J. Phys. Soc. Jap.* **7**, 107.

Ueta, M., Kondo, Y., Hirai, M., and Yoshinari, T. (1969). *J. Phys. Soc. Jap.* **26**, 1000.

van Doorn, C. Z. (1962). *Philips Res. Rep., Suppl.* No. 4.

Walker, C. T., and Pohl, R. O. (1963). *Phys. Rev.* **131**, 1433.

Warren, R. W. (1965). *Rev. Sci. Instrum.* **36**, 731.

Wertz, J. E., Auzins, P., Weeks, R. A., and Silsbee, R. M. (1957). *Phys. Rev.* **107**, 1535.

Wertz, J. E., Auzins, P., Griffiths, J. H. E., and Aton, J. W. (1959). *Discuss. Faraday Soc.* **28**, 136.

Wertz, J. E., Hall, L. C., Hegelson, J., Chao, C. C., and Pykoski, W. S. (1967). *In* "Interaction of Radiations With Solids" (A. Bishay, ed.), p. 617. Plenum, New York.

Willis, J. B. (1962). *Anal. Chem.* **34**, 614.

Witt, H. (1952). *Nachr. Akad. Wiss. Goettingen, Math.-Phys. Kl., 2A: Math.-Phys.-Chem. Abt.* No. 4, p. 17.

Yokota, R., and Nakajima, S. (1965). *Health Phys.* **11**, 241.

Zimmerman, D. W., and Jones, D. E. (1967). *Appl. Phys. Lett.* **10**, 82.

Four Basic Types of Metal Fatigue

W. A. WOOD

School of Engineering and Applied Science
George Washington University
Washington, D.C.

I. The Fatigue Problem

A. Basic Effect

The basic effect to be accounted for is that metals seem inherently incapable of withstanding repeated cycles of stress; sooner or later they develop what have become known as fatigue cracks. The effect is intriguing because they can nevertheless withstand relatively high sustained or "static" stress. For example, a typical titanium specimen might withstand a static tension of $\sigma \sim 34,000$ psi before it yields and of $\sigma \sim 54,000$ psi before it breaks; but under cycles of tension–compression $\pm\sigma$ it is likely to fracture when $\sigma \sim 18,000$ psi, little more than half its static yield and much less than its static breaking stress. Even more marked examples might be drawn from some of the structural aluminum alloys, and indeed from most of the so-called strong alloys; they are strong under static stress only, not cyclic.

Thus the basic problem might be restated in the form: what can cyclic stress do to a microstructure that static stress cannot?

B. Types of Fatigue

What the cyclic stress does do in practice depends so much on experimental conditions that it is not easy to distinguish general principles. Nor is it easy to resist the temptation of drawing general conclusions from observations that may hold only for limited experimental conditions. For example, much has been made by some authors, including this one, of fatigue cracks that originate in slip bands; and by others of cracks that grow according to "stage I" and "stage II" processes. But not all cracks originate or grow in those ways. Evidently a much needed first step in the study of fatigue, one that would clear away contradictions of the kind just noted, is to find why merely the way in which fatigue cracks form should vary. Then it might be possible to recognize specific types of fatigue, and thus to deal with the subject consistently. This is one aim of the present review.

We suggest it would be helpful to begin by distinguishing between two essentially different kinds of metal: what we shall describe as plastic metal and as elastic metal. By plastic metal we mean simple single-phase metal that in its annealed state has little or no yield strength, typical examples being aluminum, copper, or indeed any of the fcc group. These are not metals much used in engineering structures, but they are the metals that in the past have been most used for study of fatigue mechanisms. Next, to define what we mean by elastic metal, we recall that the plastic state just described is also the normal metallic state; for no normal metal, especially in single-crystal form, has a technically significant elastic range. We recall further that such a range is achieved only by introducing into a metal misfit boundaries, foreign solutes, precipitates, or indeed any obstacles that can pin the dislocations responsible for plastic flow, limit their travel under stress, and prevent them from multiplying. Thus we have a class of so-called strong metals with an extensive elastic range that is essentially a range of suppressed plastic flow. They are, however, the metals needed and used for engineering structures; their fatigue is therefore of special significance; and their obstacle-studded matrix defines a distinctive metallic state, our elastic metal.

On the basis of plastic metal in which pinning centers are weak and of elastic metal in which they are strong we can expect the following distinctive types of fatigue. Later we shall show they are in fact distinct and treat their peculiarities in more detail.

1. PLASTIC FATIGUE

We shall use this term for fatigue in plastic metal. Evidently it is fatigue by cycles of stress that can cause dislocations to move, multiply, or rearrange themselves readily. Therefore we can hardly expect the cracking to arise from any local buildup of high internal stresses or elastic strains, for these the facile dislocation and plastic flow would too easily relax. Cracking must be due rather to effects that swarms of dislocations might produce.

We shall have need to refer to two such effects. One is slip. It is known that dislocations from a dislocation source near the metal surface can escape to that surface in numbers large enough to produce an easily observed surface disturbance, the "slip band." The dislocations on their way to the surface may not keep to one lattice slip plane but, as indicated by Fig. 1a in section,

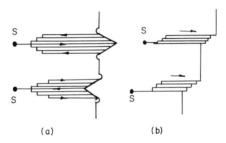

(a)   (b)

Fig. 1. (a) Band of dislocations from sources S producing surface notches or peaks because of their to-and-fro motion (cyclic slip); (b) band in unidirectional strain produces harmless surface step (static slip).

they may cross slip and climb among a band of neighboring slip planes, so that the surface disturbance is likely to have a jagged fine structure depending on how the dislocations distribute themselves in the escape band. However, providing this band remains narrow enough, the fine structure as a result of continued escape and continued to-and-fro motion of dislocations under cyclic stress could develop into a pronounced ridge or notch, and the notch could be the beginning of a surface crack. This is one effect that could hardly happen under static unidirectional stress, for then dislocations must always move in the same general direction; the most they could build up at the surface would be a harmless step as in Fig. 1b. Therefore we shall find it convenient to distinguish between "cyclic" slip bands and "static" slip bands, the first capable of the ridge-notch effect and the second not.

The other effect we shall need to refer to concerns dislocations issuing from sources within the body of the plastic metal and away from the surface. This is a tendency of the dislocations to swarm or concentrate along slip

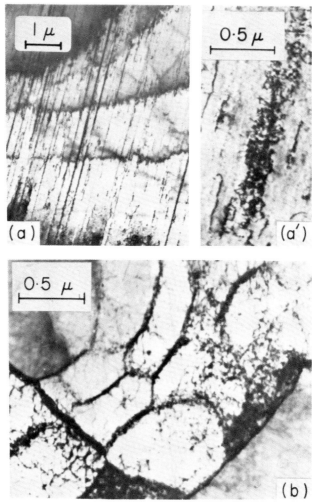

Fig. 2. (a) Lattice faults at small or F amplitudes tend to concentrate along slip zones; fine structure of one zone is shown by right inset (a'). (b) The faults at large or H amplitudes tend to concentrate along cell boundaries.

planes or along cell boundaries as in Fig. 2a,b. We shall discuss conditions for these formations later, and for indications that they are zones in which cracks prefer to nucleate or to extend; but we introduce them here because they are concentrations that again could arise easily only in the plastic type of matrix where dislocations can move around without much hindrance.

2. ELASTIC FATIGUE

We shall use this term for fatigue in elastic metal, and fatigue by stress cycles that are confined to its nominally elastic range. We shall have to add the further qualification that under these elastic cycles the pinning centers for all practical purposes are stable. Otherwise a fatigue process might arise from collapse of both pinning system and elasticity; indeed this instability we later class as alloy fatigue.

In elastic fatigue, then, in contrast with plastic fatigue, we can no longer expect a swarming of mobile dislocations and cracking on that account. But what we can expect instead, since now few mobile dislocations exist to relax them, are local buildups of internal stresses and cracking for that reason. Internal stresses are likely because an externally applied stress necessarily deforms an obstacle-studded matrix in an inhomogeneous manner. Details of this inhomogeneity may vary but the principle underlying it might be described schematically by Fig. 3. We are looking down on a metal volume

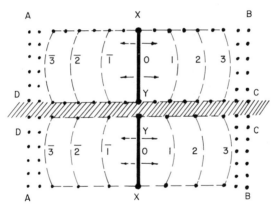

Fig. 3. Stress in regions ABCD relaxes as dislocations XY move from 0 to 3, causing stress in intervening dislocationfree layer to build up.

consisting of the two regions ABCD separated by a relatively narrow layer (shaded). The volume is bisected by the plane of the paper, which is also a slip plane containing the pinning centers and dislocation obstacles. These are represented by points, and they are so distributed that the stress indicated by arrows might easily move a dislocation XY within one of the areas ABCD, but not beyond it; for example from its initial position 0 through positions 1, 2, 3 only. Suppose each area does in fact contain a dislocation XY initially in position 0, but that the shaded layer in between is dislocationfree. Then consider the effect of applying first a static stress S. At the instant of application, $t = 0$, the stress will produce throughout the volume a wholly elastic lattice strain, one that simply stretches the interatomic spacings of the lattice. But almost at once and for a short time t, this elastic strain in the areas containing a movable dislocation will relax as the dislocation travels as far as it can, to position 3. Now while that strain is relaxing, the elastic lattice strain in the dislocationfree intervening layer is increasing, for then it alone supports the applied stress. Therefore this intervening layer is subjected to an impulsive load that might be written kSt, where k is a stress-raising factor determined by the relative sections of the dislocationfree and dislocation-containing regions. We shall present evidence that such local overloads can occur in an elastic matrix and can cause fracture.

Consider next the different effects of static and cyclic stress. Under static stress, the overload exists only for the short time t that the dislocation-containing regions are in the state of flux caused by the moving dislocation. When this flux subsides, all regions take up the applied stress by a now stable stretching of interatomic bonds; so both local overload and risk of fracture subside. But under cyclic stress, some dislocations may move to and fro continually; risk of fracture never subsides.

One point to be noticed is that the above movement of isolated dislocations can hardly be regarded as the kind of plastic strain that produces yield and slip bands because for these a large-scale escape of dislocations is needed. A second point is that the cracking is the result of elastic overstrain in a dislocationfree region; thus it is a brittle slipless rupture of atomic bonds in that region, not a ductile fracture. A final point is that the direction of cracking is therefore determined primarily by the direction of most elastic overstrain, and this in turn simply by the direction of the externally applied stress; slip directions or slip crystallography are secondary; the crack occurs essentially as it would in an amorphous matrix.

3. PLASTO-ELASTIC FATIGUE

We shall use the term plasto-elastic for fatigue again in a stable elastic matrix, but fatigue under stress cycles that now exceed its nominally elastic range and static yield. It differs from plastic fatigue, for though the cycles now produce plastic strain they do so in an obstacle-studded matrix that still prevents the dislocations responsible for this strain from moving and rearranging themselves freely. And of course it differs from elastic fatigue, for the cycles can now tear many dislocations away from their pinning centers. We can expect the plasto-elastic fatigue to show some modified features of both the plastic and elastic types: features such as attempts at plastic fracture by cyclic slip and at brittle fracture by locally inhomogeneous strain. We shall show in fact that both features may coexist, giving apparently brittle cracks in an apparently plastic matrix.

4. ALLOY FATIGUE

We shall use this term for fatigue again in the elastic type of matrix, but now in one that under cyclic stress is unstable, so that alloy fatigue is caused as much by a local rearranging of foreign atoms in the matrix as by the cyclic stress per se. We have termed it alloy fatigue because it occurs especially in metal strengthened by complex alloying, well-known examples being the precipitation-hardened aluminum alloys such as 2024 and 6075 where strength depends sensitively on precipitates in specific states of segregation; and we may expect it in any other alloy system where strength depends sensitively on particular atomic arrangements. In this way alloy fatigue illustrates one particular difficulty in studying and controlling fatigue, one arising because cracking is above all a point phenomenon, normally beginning at isolated points and so depending on conditions in the metal microstructure at isolated points. This difficulty is that we can control only the average condition of a microstructure and the average stress or strain applied to it, not the local condition at each point in it nor the local stress. Therefore the more complex the microstructure, the more likely is fatigue by inhomogeneous deformation.

5. SPECIAL CASES

These are cases where some experimental factor or some particular stage in the above major types of fatigue has attracted intensive study and led almost to a separate subject. The main ones might be summarized as follows.

a. Corrosion Fatigue. In this the cyclic stress, applied to metal in an environment that reacts with it, usually aggravates both the corrosion and the fatigue; it may do so, for example, by continually rupturing protective films at grain boundaries, slip bands, or cracks, and in that way continually exposing nascent reactive material.

We might include with corrosion fatigue the related case of "frettage fatigue" or "frettage corrosion," arising when cyclic stress may cause metal parts to rub slightly against each other, as bolted or riveted parts may. Then cyclic stress aggravates all three: local wear, local fatigue, and local corrosion.

b. Ultrasonic Fatigue. Except for special time-dependent cases, like fatigue accompanied by corrosion or frettage, a given fatigue mechanism in general does not significantly alter over the range of cycle frequencies up to some 200 Hz, the "low frequency" or LF range in which most fatigue testing hitherto has been carried out; but it may at higher frequencies, particularly at ultrasonic frequencies of some 20,000 Hz or more. Thus ultrasonic or HF fatigue, as we hope to show, is of theoretical as well as technical interest.

c. Fatigue by Crack Growth. Study of how cracks grow under cyclic stress, whether cracks initiated by fatigue or in any other way, has led to empirical and theoretical formulas for crack growth; and sometimes to claims that all fatigue might be treated for practical purposes as fatigue by crack growth. A crack growth or fatigue process that is formulated in mathematical terms is clearly of special value in engineering design. But the physical interpretation is of more interest here, and we shall discuss rather various experimental observations of how cracks begin to grow in the major types of fatigue defined above, showing for example that they grow differently according to whether fatigue is plastic or elastic.

d. Accidental Fatigue. By this is meant fatigue started by some accident of structure, a common example being the "metallurgical crack" where a foreign inclusion may cohere imperfectly with the metal matrix. It is not a process that can be treated systematically, except possibly by the formulas of crack growth, but it is included here for two reasons. The first is that it is common especially in elastic metal, for this has not the capacity of plastic metal to damp off an incipient crack by easy plastic flow. The second is that because elastic metals are so accident-prone and because they so often fail accidentally, it has been widely assumed that all elastic fatigue must be accidental fatigue. We shall hope to show that this assumption is too sweeping.

II. Experimental Study

We may expect each type of fatigue to develop in some primary characteristic way, but to show secondary modifications according to experimental variables like stress cycle, surface preparation, or test temperature. So a desirable second step is to produce the fatigue first under the simplest possible experimental conditions, so obtaining a standard of comparison, and then to introduce variables one by one. It may be impracticable to cover all variables; it certainly cannot be done in this review; but it should be possible from experience with selected variables to anticipate the effect of others. Here, where possible, we shall take as a standard of comparison metal that is initially annealed, polycrystalline, electropolished, and subjected to cycles of fixed stress or strain symmetrically alternating about zero. Annealing removes the variable of prior internal strain; polycrystalline material is more reproducible than single-crystal; electropolishing provides a workfree featureless surface suitable for microscopy; and the symmetrical cycle does not superpose unidirectional deformation on the cyclic type. We shall also assume that the standard is stable enough to be tested in air and at room temperature.

What form the stress cycle takes, whether alternating push–pull or torsion or bending, is immaterial; but a simple flexible arrangement is an alternating twist $\pm\theta$, especially one permitting steady axial tension σ to be superposed when desired as in Fig. 4, which also indicates essential dimensions of

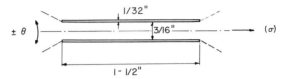

Fig. 4. Test arrangement used for text illustrations. If desired, axial tension σ can be superposed on alternating twist $\pm\theta$.

specimens used for several later illustrations. No brief is held for this particular method, but it has some advantages in basic research. One is that since shear is generally responsible for starting plastic flow and tension for starting crack growth it is useful to have facilities for varying torsion and tension independently. A second advantage is that the arrangement can test how stable the pinning of dislocations may be under the cyclic torsion, for dislocations that are free or are easily unpinned by the torsion will drift axially under the tension and start an axial creep. Applications of this kind also are touched upon.

Whatever the mode of stressing, the metal in basic studies has to be examined by some method that permits inferences about how the dislocations and other lattice defects responsible for mechanical properties have behaved. The methods most used have been mechanical testing, microscopy, and X-ray diffraction. Mechanical tests, mainly of hardening and hysteresis loops, have shown in particular that behavior under cyclic stress differs significantly from behavior under static stress; but the behavior measured is that averaged over the whole bulk metal, when of special significance might be behavior in the surface layer only. Microscopy and X-ray diffraction go to the other extreme by showing surface behavior only, either of the original surface or of an etched section. So a problem in reviewing basic work on fatigue is to reconcile different types of observation as well as the various types of fatigue.

A. Mechanical Studies

Mechanical studies have led to well-known generalizations about dislocation behavior under cyclic strain, but mainly about their behavior in plastic metal. It will be convenient first to summarize these generalizations and then to indicate how they break down when metal is not plastic.

1. MECHANICAL TESTS ON PLASTIC METAL

Main generalizations may be described by reference to the hysteresis loops in Fig. 5 for the typical plastic metal copper under an alternating twist $\pm\theta$. (The specimen in this test was also under axial tension, used to illustrate a later point, but the tension was too small to interfere significantly with the features under discussion.)

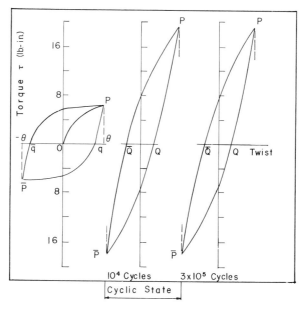

Fig. 5. Initial hysteresis loop of copper under alternating twist $\pm\theta$, and loops in its cyclic state.

First, peaks P and $\underline{P}$ of the hysteresis loops initially increase with cycles; so the metal is said to exhibit cyclic strain hardening.

Second, a loop under an amplitude that is not too large, and which therefore permits many cycles before fatigue cracking, begins after enough cycles to repeat itself; the metal is said to have reached a cyclic state or a saturated hardness. This hardness level and the number of cycles needed to establish it at a particular amplitude follow the scheme indicated by Fig. 6. At very large amplitudes H the level is high and barely reached before fracture; at small amplitudes F it is lower and reached sooner; and at still smaller amplitudes S it is very low and reached very soon.

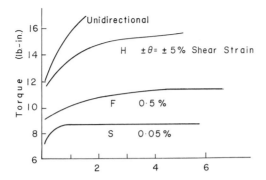

Algebraic Sum Of Reversed Shear Strains θ (%)

Fig. 6. Cyclic hardening curves for aluminum under H, F, S amplitudes of alternating twist $\pm\theta$.

Third, a cyclic state implies that a plastic amplitude $Q\bar{Q}$ in Fig. 5, reverses continually without significantly disturbing the dislocation arrangement responsible for saturation hardening. We then have a reversible plastic strain analogous to a reversible elastic strain.

Next we recall the kind of dislocation model needed to account for these features. Plastic strain is due mainly to movement of dislocations through the metal lattice, and strain hardening to growing interference with this movement. Interference may arise from increasingly strong reactions between strain fields of multiplying dislocations, increasingly complex shapes of individual dislocations, and an increasingly imperfect crystal structure of individual grains as different regions deform by different amounts and become differently oriented. There is as yet no definitive treatment of these factors, but details need not concern us here. The essential point is that an otherwise mobile dislocation in the saturation-hardened metal must be pinned in a way that permits the two particular movements indicated by Fig. 7a–d. Here the pinning centers, whatever they may be physically, are depicted as points temporarily holding the dislocation XY at the points it is shown passing through.

Then one permitted movement must be an oscillation of the dislocation about this temporary pinned position, indicated in Fig. 7a by a bowing of the dislocation between its pinning centers. This movement is needed to account for the hysteresis loop occurring when a metal is simply unloaded and reloaded as in Fig. 7b.

The other permitted movement is a larger reversible one corresponding to the reversible plastic strain $Q\bar{Q}$. Suppose the dislocation is first in the position Q of Fig. 7c, a position corresponding to the unloaded state at Q in the hysteresis loop of Fig. 7d. Then to account for the larger movement, the

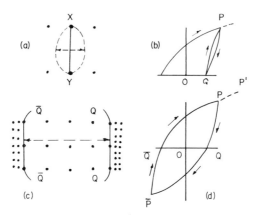

Fig. 7. Dislocations XY oscillating about pinning points in (a) correspond to small loop (b). Dislocations moving between bounds QQ and Q̄Q̄ in (c) correspond to wide loop (d).

reverse load in direction QP̄ must tear the dislocation from its pinning centers. Now to do this in practice needs a much smaller load than it does to continue deformation in the direction Q̄PP'; plastic metal has no reverse yield strength (Bauschinger effect). Therefore, as also indicated by Fig. 7c, pinning centers must be distributed unevenly, being crowded together in the direction of difficult motion. The same conditions repeat themselves when the specimen is unloaded from P̄ to Q̄. Therefore the larger movement QQ is one between bounds set by zones of crowded pinning centers or, since these themselves are likely to attract lattice defects, between zones of crowded defects. Somehow the dislocation, as it moves to and fro under the cyclic strain, must pile or push some pinning centers or obstacles before it and deposit them at the limits of its path. Evidence of such zones appeared in Fig. 2; other evidence will appear later when microscope studies are discussed more fully.

This kind of model raises the question: what relation can it possibly have with cracking? For there seems no reason why dislocations that simply oscillate about pinning centers as in Fig. 7a or between bounds as in Fig. 7c should ever start cracks; the movements are reversible. Thus a prolonged cyclic state should correspond with a prolonged fatigue life. This in fact it does. Consider the typical s/n curve for plastic metal in Fig. 8, n being the number of cycles to fracture at amplitude s. The ranges there marked H, F, S correspond to the ranges H, F, S in Fig. 6. Thus the short-life range H corresponds to amplitudes that can barely produce a cyclic state before fracture, whereas the long-life ranges F and S correspond to amplitudes that produce a cyclic state rapidly and maintain the metal in that state for most of its life.

It seems to follow that cracking must originate in regions of a metal where

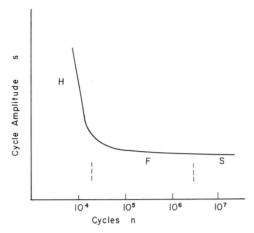

Fig. 8. Typical H, F, S ranges in schematic *s/n* curve for plastic metal.

unstable conditions can persist even at F and S amplitudes. They must also be regions that can escape detection by bulk measurements like those on hardness or hysteresis loops. There are two such regions: the surface layers of the metal and the internal zones of crowded lattice defects indicated by the model of Fig. 7. These then will call for special study by microscopy.

2. MECHANICAL TESTS ON ELASTIC METAL

We shall have to consider separately conditions (a) where stress cycles are within the elastic range, (b) where they exceed it, and (c) where they make it unstable.

a. Elastic Cycles. One immediate difference between behavior of plastic metal and stable elastic metal is shown by their *s/n* curves. The curve for plastic metal is continuous; the H, F, S ranges of Fig. 8, for example, pass smoothly from one to the other. But the curve for stable elastic metal at a particular elastic amplitude shows a discontinuity; fatigue lives above this amplitude are relatively short and below it indefinitely long; the amplitude is said to define a safe fatigue limit or safe range. Figure 9 illustrates this safe range for titanium tested in alternating torsion under the conditions of Fig. 4; a safe limit occurs at the amplitude $\theta \sim 0.003$ shear strain.

The effect at once follows from the model of elastic metal: it simply means that up to the safe limit all the amplitude can do is merely make dislocations oscillate about their strong pinning centers; it cannot tear them away and start irregular deformation. Two points of interest arise. The first is that this safe fatigue range should measure how stable the static elastic range of the metal may be. If the two are of the same order, then the static elastic range

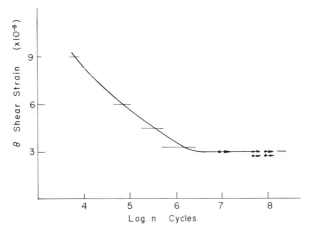

Fig. 9. Safe limit about $\theta = 0.003$ shear for titanium in alternating torsion.

must be highly stable; dislocations must remain strongly pinned throughout the range. Simple carbon steels, where elasticity is achieved by strain aging and by finely dispersed carbides, exemplify high stability; their safe limit is often little lower than their elastic limit. Titanium, where a sensibly linear elastic range is achieved by strain aging, can provide an example of less stability. Thus the titanium used to obtain the s/n curve of Fig. 9 gave the static torque/twist curve shown in Fig. 10. It will be seen that the linear static

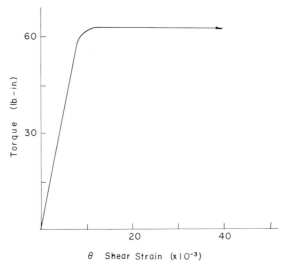

Fig. 10. Yield at about $\theta = 0.009$ shear for titanium in unidirectional torsion (dead loading).

elasticity holds until the shear strain θ approaches 0.009; the safe limit of 0.003 is only one-third of this.

The second point is: how then does cracking occur at amplitudes between the safe fatigue limit and a much higher static elastic limit? The general answer hitherto has been: by accident. It has been supposed that in titanium, for example, this fatigue cracking is the result of structural accidents like occasional weak grains that may fail by plastic fatigue. However, a more systematic cause might be postulated. It is reasonable to suppose that between the safe limit where dislocations cannot escape from their pinning centers and the elastic limit where they begin to escape in large numbers there exists an intermediate condition where isolated dislocations can make a limited escape from occasional centers in the manner described by the earlier Fig. 3. A test would be that the resulting fracture, as also noted in the earlier discussion, should be by a brittle slipless shear crack. This in fact is what can occur (see Fig. 30).

b. Plasto-Elastic Cycles. These again produce deviations from the generalizations based on plastic metal. They are well brought out by stress/strain curves or hysteresis loops obtained by dead loading in torsion; the dead-loading allows any plastic strain to proceed to completion, and the torsion allows it to proceed in its natural form of pure shear. Differences between plastic and elastic metal then appear even in their response to simple unidirectional deformation. Plastic deformation of the plastic metal under each dead load soon comes to a virtual standstill; the stress/strain curve has a meaning. But that of elastic metal may be surprisingly prolonged; in titanium or chromium, for example, it may go on for days, so much so that their plastic behavior can be represented only by a family of creep curves; their stress/strain curves have little meaning (Wood, 1971). Apparently when dislocations escape from the strong pinning centers in these elastic metals as yield is exceeded, they cannot go far before they are captured by other centers; then they must begin their escaping again; plastic deformation becomes a prolonged irregular stop–go process.

This difference of elastic from plastic metals is carried over into their cyclic strain hardening. A plastic metal like copper under cyclic strain begins hardening rapidly and systematically as illustrated by Fig. 11. But an elastic metal like titanium may hardly harden at all, even, as shown by Fig. 12, under amplitudes well in excess of its yield point. Thus generalizations about cyclic strain hardening have meaning only for plastic metals, not elastic.

c. Unstable Elasticity. Evidently the generalizations are likely to have less meaning still for elastic metals that become unstable under cyclic strain. Then the cyclic strain instead of hardening may cause softening. An example is given in Fig. 13.

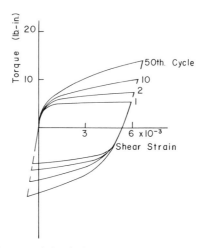

Fig. 11. Rapid systematic hardening of plastic metal in cyclic torsion (Cu).

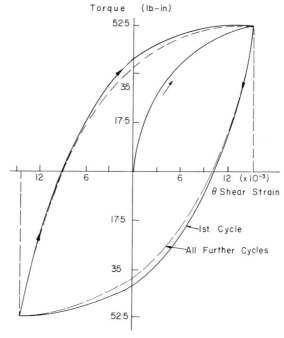

Fig. 12. Slow inefficient hardening of elastic-type metal in cyclic torsion (Ti).

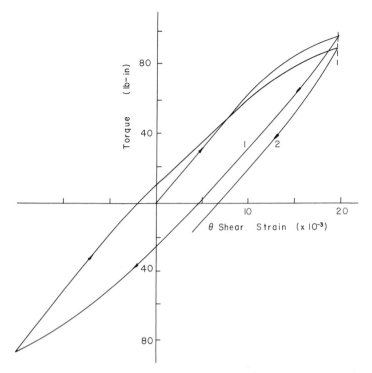

Fig. 13. Softening of complex alloy in cyclic torsion (hardened Cr/Mo steel).

Thus we conclude these sections on mechanical studies by making the point that they permit useful inferences about how dislocations respond to cyclic strain; that, however, they can cause confusion unless due regard is paid to the kind of metal under test; but that much of this confusion disappears if distinctions are made in particular between plastic and elastic metal. Next we show that a similar situation arises in making inferences from microscope studies.

B. Microscope and X-Ray Studies

To facilitate comparisons, the micrographs used here for illustration are from specimens tested by the arrangement of Fig. 4. Most are scanning electron-micrographs (SEM) directly from the electropolished surface of the specimen or from a section, for the SEM technique clarifies many of the earlier observations by optical microscopy; in any event relevant optical observations have been recently discussed elsewhere (Wood, 1971). Both the SEM and the occasional optical micrographs (opt) are similarly mounted with their horizontal direction parallel to the specimen axis, so that comparisons can be made too between crack directions and stress directions.

1. PLASTIC FATIGUE

We begin with microscope studies of plastic fatigue because again they have been the basis of generalizations that in fact hold only for special cases of fatigue. We consider first how the studies may clarify some of the problems raised by the mechanical tests, but not solvable by those or other tests that measure only properties in bulk.

a. Regional Strains. When we fix a bulk strain amplitude, what in fact we fix is the sum of local strain amplitudes that may vary from one region to another in each metal grain. So one crucial question is, what are the regional strains; for it is these, after all, that determine fracture. So far, unfortunately, no technique has been devised to measure them. But at least it is possible to tell whether they stay constant during successive cycles, and since this information itself is useful, it is noted first.

The information follows from tests by X-ray diffraction. Classic metallographic studies during the 1920s and 1930s showed that if a metal crystal is constrained to extend plastically in a given direction, it has to tilt its slip planes as in Fig. 14a,b. Amount of tilt depends on amount of extension. So a

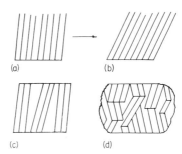

Fig. 14. (a) Lattice planes of crystal continually extending in given direction must tilt as in (b). They must develop different tilts as in (c) if different regions extend differently. A grain in polycrystalline metal thus becomes inevitably disoriented as in (d).

crystal extending by different amounts along its length develops variously tilted regions with more or less well-defined boundaries of lattice faults as in 14c. And in general, any unevenly deforming crystal or grain must become similarly " disoriented " as in 14d. This disorientation the X rays detect sensitively, for reflecting X rays from the lattice planes of a grain is in this respect like reflecting light from a mirror; if the mirror breaks into a mosaic of differently tilted parts, its reflections spread.

The test is significant here because if a bulk plastic amplitude progressively disorients a metal, it must be made up of regional strain amplitudes that progressively vary, for only by varying can they continually alter the orientation of their regions. Conversely, when the bulk amplitude produces no disorientation, or no progressive disorientation, it must be made up of regional strain amplitudes that are constant. Applied to initially annealed metal, where lattice orientation in each grain is at first sensibly uniform, the test shows that an alternating plastic strain breaks down a grain into disoriented regions only during the first phase of cycling while the metal is strain hardening in the manner indicated earlier by Fig. 6; and that during the second phase, when hardening ceases and the metal attains a cyclic state, further disorientation stops (Wood and Segall, 1957). This result is of interest, first, because it leads at once to a distinction between large or H bulk amplitudes and small F or S amplitudes. For, as also indicated by Fig. 6, at an H amplitude the first phase of hardening (and disorientation) takes up most of the fatigue life; whereas at a small F or S amplitude it is the second phase of a cyclic state that takes up most of the life. Therefore the large H amplitudes for most of the life are made up of regional amplitudes that are in a state of flux, while the small F or S amplitudes are made up of regional amplitudes that for the most part are stable.

The result is of interest, secondly, because regional amplitudes can be in a state of flux only if the forward bulk strain during the forward half of a cycle distributes itself in one way among the various regions, and the backward bulk strain during the backward half in a different way; for example, though possibly an extreme case, the forward strain might distribute itself in one set of regions and the backward strain in quite another set as depicted by Fig.15a. This kind of condition, then, characterizes an H amplitude. In contrast the condition ceases to hold if the forward strain in any one region becomes the backward strain in the same region as in Fig. 15b, for that region can no longer exhibit progressive disorientation. So this local reversibility of regional strain characterizes a small F or S amplitude.

b. Surface Abnormalities. The above conditions of locally variable or locally reversible strain, as may be shown by X-ray tests on sections of a cycled metal, characterize all regions throughout its volume. But they should also be reflected to some extent in the way the metal surface behaves, where they can be studied by microscopy.

For example, if H amplitudes are in fact made up of regional strains that need not reverse precisely, they should produce virtually the same kind of slip in the surface as static or unidirectional deformation does, differing only in that this static slip should strongly rumple an initially flat surface for reasons evident from Fig. 15c. This the microscope at once confirms. Thus

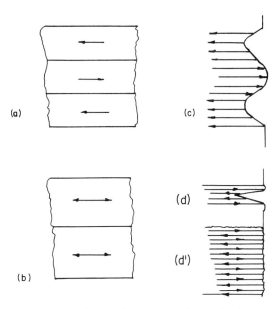

Fig. 15. (a) H amplitudes tend to produce nonreversing regional strains, causing surface rumpling as in (c). (b) F and S amplitudes produce reversing regional strains, causing concentrated surface disturbance as in (d) or dispersed disturbance as in (d′).

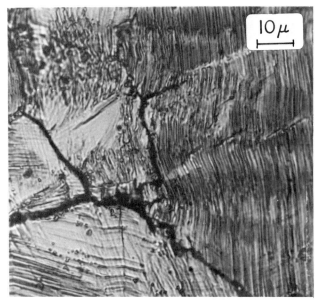

Fig. 16. H amplitude produces irregular static slip and irregular boundary cracks (alternating torsion, 70/30 brass, opt).

Fig.16, typical of an H amplitude, shows simply static slip bands that are irregularly curved as a result of rumpling and irregularly distributed as a result of generally uneven deformation; moreover it shows that when surface cracks appear these too are irregular in shape.

In contrast, the small F or S amplitudes should give rise to what we have termed cyclic slip, where slip movements during successive cycles go back and forth in the same region. Then if they remain in a narrow band as in Fig. 15d, they can build up a concentrated surface disturbance and even a slip-band fissure. This again the microscope confirms. Figure 17 shows typi-

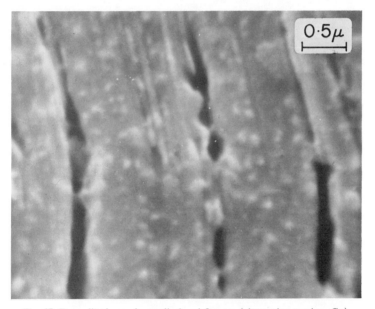

Fig. 17. F amplitude produces slip-band fissures (alternating torsion, Cu).

cal slip-band fissures built up by an F amplitude. If, however, the to-and-fro slip movements do not remain in a narrow band but spread by cross slip or climb into a wide band as in Fig. 15d′, they can build up in the metal surface no more than a harmless fine ripple. Spreading of this kind is illustrated first by Fig. 18, where the slip has remained sufficiently concentrated to produce at least some fissures but has begun to spread by clearly visible cross slip; then by Fig. 19 where the slip, because now it must have spread faster than it could form fissures, has produced a barely resolvable harmless dispersion. This last condition is typical of an S amplitude.

Thus the microscope introduces extra details about the reversible strain characterizing small amplitudes by showing how its effectiveness in fatigue

Fig. 18. Slip bands at F/S amplitude beginning spread by cross slip (alternating torsion, Cu).

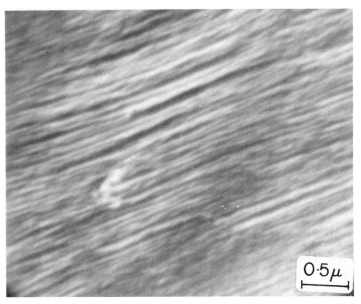

Fig. 19. S amplitude produces harmless spread of slip (alternating torsion, Cu).

depends on how concentrated it may be. It is not difficult to account for main details. Dislocations issuing from a source and producing the observed surface disturbances as in Fig. 15d are likely to form first a narrow band and then a wider and wider band as cycles continue; for some of the dislocations issuing from the source in one cycle may not reach the surface in that cycle, and these are likely to form an atmosphere of dislocations that strain hardens the neighborhood and forces newly issuing dislocations to cross slip or climb into adjacent slip planes. So conflict arises between the tendency of the dislocations to disperse and their chance to build up a concentrated surface disturbance or fissure. It is reasonable that fissuring should prevail at an F amplitude rather than at the smaller S amplitude. The larger amplitude causes the source to emit a larger burst of dislocations in each cycle than a smaller amplitude does, and therefore more dislocations will reach the surface in each large-amplitude cycle. So the chance of the larger amplitude building up a fissure before cross slip or climb interferes is higher. At the same time, factors that facilitate cross slip or climb, such as elevated temperature, should prevent any fissuring at all. In this way the argument can be tested by experiments; so far the results confirm it (Wood, 1971).

   c. *Internal Abnormalities.* The spectacular surface disturbances have tended to distract attention from the to-and-fro dislocation movements that must extend also throughout the metal interior, and from the possibility that these too may build up abnormalities of some kind. Indeed, as noted when discussing the Bauschinger effect, it seems necessary that they should at least build up concentrations of lattice faults at the limits of their to-and-fro paths (Fig. 7c).

   Internal concentrations have in fact been detected. In one method, a classic metallographic one, the polished section of a cycled specimen is etched with a reagent known to attack preferentially any zones where lattice faults may have concentrated. These researches have shown that in a specimen cycled at H amplitudes the faults tend to form and concentrate inside a grain at cell or subgrain boundaries, but that in one cycled at F amplitudes they concentrate especially in straight zones along slip directions inside a grain. The H amplitude result is to be expected because of the disorientation that also characterizes large amplitudes, and the F amplitude result because of the slip concentrations characterizing F amplitudes. Both are illustrated by Figs. 20 and 21.

   This method has the advantage that the etching should not disturb the distribution of lattice faults but the disadvantage that it cannot identify them. However, identification is possible by the second method, that of electron transmission through a thin foil extracted from the section surface. Unfortunately, extraction and thinning of the foil, especially one of plastic

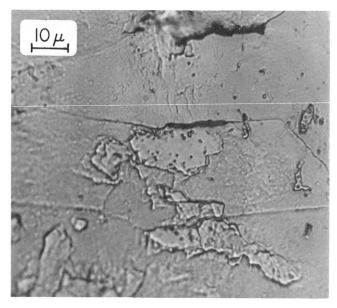

Fig. 20. Etched-up cell boundaries on sectioned metal after H amplitude (alternating torsion, Cu; note boundary ruptures).

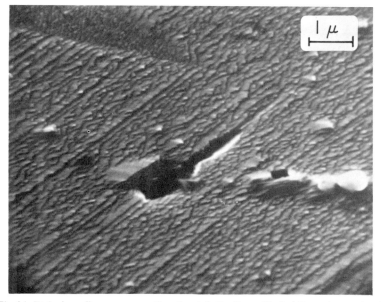

Fig. 21. Etched-up slip zones on sectioned metal after F amplitude (alternating torsion, Cu; note point ruptures).

metal, may disturb the fault distribution, but it is not impossible to conduct the etch technique and transmission technique on the same sections and obtain corresponding distributions. This was done for the results illustrated by Fig. 2 (Grosskreutz *et al.*, 1966). These indicate that both the internal cell boundaries and the internal slip zones are concentrations primarily of dislocations, dipoles, loops, and their debris, with dislocation densities ρ reaching orders 10^9 and 10^{10}.

d. Point Rupture and Fissures. It is reasonable to expect cracking to begin where the above local deformations or local distortions of the metal are abnormal. The foregoing observations indicate three localities in particular: (a) surface grain and subgrain boundaries, especially at H amplitudes; (b) surface disturbances where cyclic slip concentrates within a grain, especially at F amplitudes; and (c) internal surfaces and zones where lattice faults concentrate abnormally, boundary surfaces at H amplitudes and slip zones at F amplitudes.

Consider first the internal effects. Optical-microscope studies made some years ago on etched sections indicated that cracks could in fact nucleate preferentially in the abnormal zones within a metal, and that they began as point ruptures which could multiply and coalesce as cycling proceeded

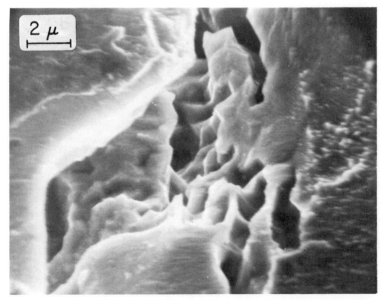

Fig. 22. Cell boundary on etched section showing point-by-point origin of rupture (alternating torsion, H amplitude, Cu).

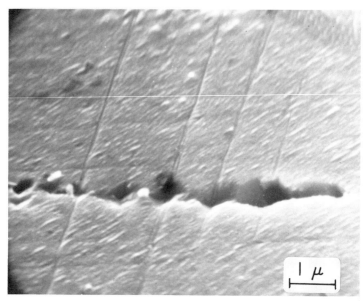

Fig. 23. Slip zone on etched section showing slip-zone rupture (alternating torsion, F amplitude, Cu).

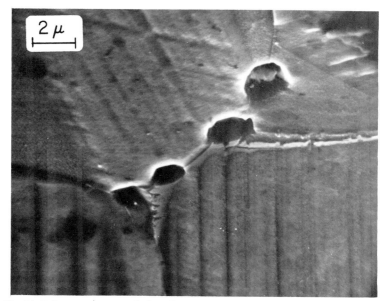

Fig. 24. Point ruptures on etched section at meeting of internal slip zones and boundaries (alternating torsion, Cu; note starting boundary crack).

(Wood, 1971). Later studies by scanning electronmicroscope provide confirmation. Figure 20 showed minute ruptures on etched-up internal cell boundaries, and Fig. 22, showing one of these ruptures at high magnification, brings out its point-by-point origin. Figure 20 showed point ruptures on etched-up slip zones, and Fig. 23 shows them after they have coalesced into a longer rupture extending along a slip zone. Figure 24 shows point ruptures where internal grain boundaries and internal slip zones meet, and further confirms the general principle that preferred sites for rupture are where lattice faults tend to pile up. How ruptures arise there has been a much discussed topic of dislocation theory (e.g., Cottrell, 1959), but this need not concern us here. It is sufficient to note that theory provides sufficient reasons why ruptures could form out of piled-up lattice faults, and merely to concentrate here on experimental observations about their appearance and progress.

Consider next the surface effects. These we may expect to supersede any internal ruptures, for cracks nucleating in surface layers are in layers that deform more freely than internal regions and more freely undergo atmospheric attack. However, it is of interest to inquire first whether surface cracks too may begin as point ruptures in the abnormally deformed localities, and whether they too may grow by coalescence.

It is not as easy to detect first signs of abnormal points in the surface as it is in the interior, for small effects tend to be swamped by general surface rumpling and much harmless slip. But evidence can be found by using for the test a metal like copper that reacts mildly with air, for the air then acts as an etchant during cyclic strain and by attacking overactive points preferentially can reveal them before they are swamped. Figure 25 shows them for an F amplitude. The attacked active centers are clear, spaced with typical regularity at intervals of 10^{-4}–10^{-5} cm along incipient slip bands. These centers appear even more clearly if the test temperature is slightly raised; since they react as if during the cyclic strain they are hotter than adjacent metal, they might appropriately be termed hot spots. Other examples occur in later illustrations.

It is not difficult to demonstrate the further point that the hot spots become point ruptures that extend mainly by multiplying and coalescing along an abnormally deformed zone. But a matter of technique arises. Small ruptures during cyclic strain commonly extrude much debris, a much studied effect attributable to frettage corrosion between the rupture walls. The literature contains many illustrations, and there is hardly need to add to them here. Instead it is desired to make a point often overlooked, namely, that the extrusions in one sense are a nuisance, for they obscure the ruptures from which they issue, and that it is often more informative to remove them. This can usually be done by a light etch that dissolves the debris without

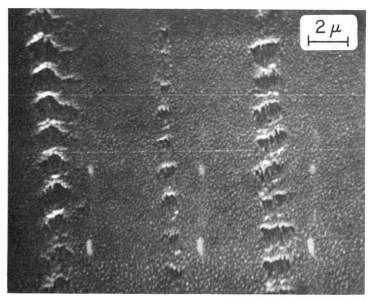

Fig. 25. Hot spots along incipient slip bands (F amplitude, alternating torsion, Cu).

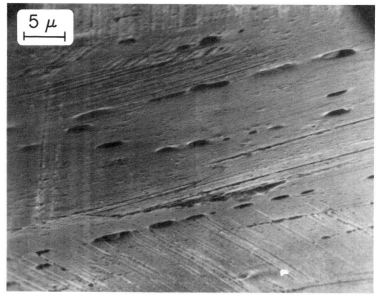

Fig. 26a. Point ruptures in surface slip bands (F amplitude, alternating torsion, 70/30 brass, light etch).

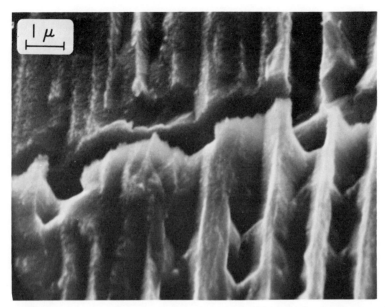

Fig. 26b. Point ruptures coalescing into slip band fissure (F amplitude, alternating torsion, Cu, light etch).

significantly attacking the metal at large. When this is done, the point ruptures and their multiplication are generally observable. Figure 26a shows early point ruptures in slip bands, and Fig. 26b shows that a point-by-point origin can be discerned in later stages when slip bands become virtually continuous ruptures. Other examples appear later (Figs. 37 and 38). Similar effects are observable along boundaries.

The elongated ruptures formed by point ruptures coalescing along slip bands, and so typical of F amplitudes, are conveniently termed slip-band fissures. Those forming along grain or subgrain boundaries, since these arise from abnormal deformation due to lattice misfits, we may term misfit fissures. It will be noted that we are reserving the term fissures for ruptures within or shorter than a grain. Various fissures appeared in Figs. 16, 17, 23, and 26. Figure 27 shows how F amplitudes in particular can produce fissures by the score even in individual grains.

e. Linkage Cracks. The next step in the crack process is not so much by growth of individual fissures as by their linking across grain boundaries to form what might be termed linkage cracks. Apparently, individual fissures do not self-propagate readily, an inference following from the feature shown, for example, by the last illustration that they can form in large numbers even

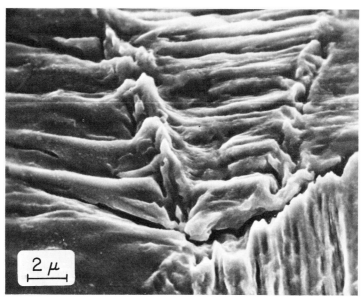

Fig. 27. Typical surface after F amplitude, showing multiple slip band and misfit fissures and abnormal boundary deformation (alternating torsion, Cu, light etch).

in one grain long before final failure of a specimen. It seems that the fissures can achieve a more effective *de facto* propagation by linking.

The question then arises whether they link in any preferred direction. At large or H amplitudes, where fissures form mainly at misfit boundaries, the linkage has little choice but to follow an irregular course from boundary to boundary. In practice this course, especially in large-grained specimens, seems to have no special trend. Whether it might have a trend in fine-grained specimens, where the linkage crack has more choices, has not been studied. Of perhaps more interest are conditions at small or F amplitudes where the innumerable slip-band fissures as well as misfit fissures present the linkage crack with a multiplicity of choices. Here there is much evidence that the linkage crack clearly develops a trend in a direction of maximum tension. The crack in Fig. 28 is typical. The horizontal direction of this photograph is parallel to the axis of a specimen tested in alternating torsion by the arrangement of Fig. 4, so the 45° trend of the crack is a direction of maximum tension. By using the arrangement of Fig. 4 it is possible to demonstrate the apparent reason for this trend. If an axial tension is superposed on the alternating torsion, it pulls open those fissures at right angles to it. So the tensile crack is merely following those fissures that tend to open up most in the alternating-torsion test.

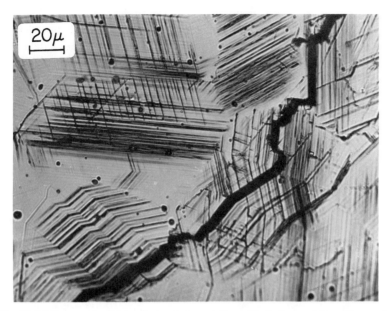

Fig. 28. Linkage crack with tensile trend through F amplitude fissures (alternating torsion, 70/30 brass, opt).

This observation that linkage takes a tensile trend is of special interest because it shows that two commonly stated rules in the literature about the directions of cracking are overgeneralizations. These are the rules that cracking follows what have been termed stage 1 and stage 2 directions (e.g., Forsythe, 1962). In stage 1 the crack as it first forms is supposed to extend along a direction of maximum shear, the reason given being that it is virtually a linkage of slip-band fissures from grain to grain. This rule might hold in special cases where the only fissures produced by fatigue might be slip-band fissures, for it is known that these always follow that crystallographic slip direction in a grain along which the shear stress is a maximum. But the rule can hardly hold here for the plastic fatigue under discussion. In the first place, at H amplitudes there are no slip-band fissures for a crack to link up, only randomly directed misfit fissures. In the second place, at F amplitudes, where slip-band fissures do occur, there also occur competing misfit fissures, and the general trend of the crack if anything is in a tensile direction, not one of shear. A similar overgeneralizing marks the stage 2 rule. According to this, the stage 1 shear crack turns in stage 2 to a direction of maximum tension when it becomes large enough to overide the crystallographic fissures. So stage 1 is sometimes termed the crystallographic stage and stage 2 the noncrystallographic. Evidently this rule too can hardly hold here, for the tensile linkage of fissures shown is entirely crystallographic. It will appear later, however, that there are indeed types of fatigue where the stage 1 and 2 rules do hold, but that they are types based not on behavior of plastic metal but of what we have described as elastic metal.

f. The Terminal Crack. The linkage cracks do not themselves constitute the final stage of failure in plastic fatigue. Since many fissures form in many parts of a specimen, so also do many linkage cracks. But out of these in practice emerges one long enough to propagate itself and supersede the rest, and this becomes the terminal crack. The longer crack may occur because one linkage crack joins another one directly, or because linkage cracks approach each other and intensify the fatigue deformation between them, then joining through subsidiary cracking of the intervening region. This effect is illustrated by Fig. 29. The final course of this terminal crack in the

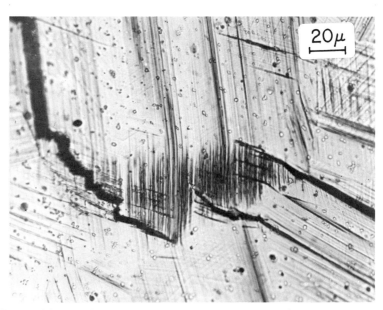

Fig. 29. Linkage cracks causing abnormal slip in intervening zone (alternating torsion, 70/30 brass, F amplitude, opt).

small specimens used normally for plastic fatigue testing is simply the direction that tends first to encircle the specimen, and this, for both tensile and torsion testing, is generally the transverse direction.

We conclude this discussion by noting that the basic plastic fatigue exhibited by metal tested under the simplified conditions chosen as a standard of comparison also provides a basis for interpreting behavior when the standard test conditions are modified. The following are some illustrations taken from more detailed descriptions (Wood, 1971). The basic observations discussed in the preceding sections show that both grain and subgrain misfit boundaries provide major sites for ruptures in plastic fatigue; slip bands are significant only in the special case of cyclic slip at F amplitudes. Therefore control of boundaries permits some control of cracking. So one good measure, if it were always practicable, would be to eliminate them; and in fact several observations show that fatigue resistance of single-crystal metal is superior to that of polycrystalline metal. Another measure, more practicable, is to align the boundaries in directions along which boundary fissures and their linkages might do least harm. This can be done by appropriate cold-working of the metal or by suitably cooling from the melt. Similarly, some knowledgable control can be exerted over the slip-band fissures typical of F amplitudes. To suppress them it is only necessary to cold-work the metal sufficiently; then the resulting disorientation and multiplying of dislocation sources prevents even F amplitudes from producing the cyclic type of slip. To exaggerate them it is necessary only first to produce the slip-band fissures by an F amplitude and then subject the metal to an H amplitude. Then the larger tensions inherent in the larger amplitudes pull the fissures wide open in a striking fashion. Thus, as illustrated by these examples, it is possible by making appropriate allowances for modifications in the standard experimental conditions also to resolve many of the apparent contradictions arising in the literature because different investigators have used different experimental conditions.

2. ELASTIC FATIGUE

With plastic fatigue described above in enough detail to provide a standard of comparison, relatively brief treatments of the other types of fatigue will suffice to show their distinctive peculiarities. Those of elastic fatigue may be conveniently illustrated by the annealed titanium also used for the mechanical tests of Figs. 9 and 10. This material has the advantage too that its structure is simple, for it owes its elastic range to a strain aging due to no more than normal commercial impurities dispersed on an atomic scale. Though most of its grains in this range behave elastically, in the sense that they exhibit no visible slip or other signs of plasticity, certainly none comparable with those of plastic fatigue, yet a few of its grains in general do exist in a weaker condition and do exhibit some slip. But the presence of these weaker grains here has some advantages, for while elastic fatigue of the elastic grains is going on they are developing plasto-elastic fatigue under identical conditions, a point that will be utilized. Again to facilitate comparisons, all reproduced photomicrographs are mounted with their horizontal direction parallel to the axis of specimens tested by the arrangement of Fig. 4. So horizontal and vertical directions in the micrographs correspond to directions of maximum shear in the specimen, and 45° directions in the micrographs to directions of maximum tension. Elastic amplitudes then produce the following unmistakeable differences from plastic fatigue (e.g., MacDonald and Wood, 1972).

a. Slipless Shear Cracking. First the elastic metal can develop transgranular cracks across grains that show not the slightest sign of slip; Fig. 30 is a typical example. Second, this crack can then propagate across adjacent grains that still develop no slip; this Fig. 31 exemplifies. Finally, as both illustrations show, the crack forms and extends always in a general direction of maximum shear, here the horizontal one. So it now obeys the stage 1 rule.

Thus one immediate peculiarity is that a crack can form and grow without any of the massive slip needed to start cracks in plastic fatigue; and another that the crack follows the stage 1 rule of shear without any of the slip-band fissures that that stage is said to require. Neither is this stage 1 crystallographic, as it is supposed to be, for a crack like that in Fig. 31 goes on its way through one slipless grain after another without being diverted significantly by their various crystallographic orientations. On this evidence it seems,

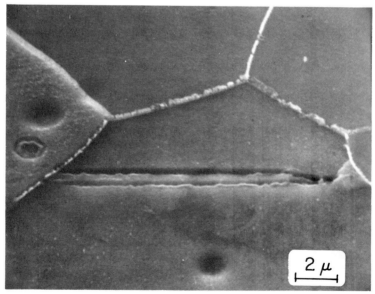

Fig. 30. Slipless transgranular shear crack in elastic metal under elastic amplitude (alternating torsion, Ti).

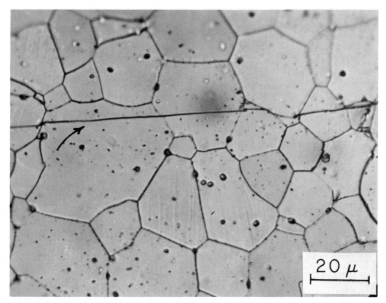

Fig. 31. Slipless shear crack unaffected by different orientations of traversed grains (alternating torsion, elastic amplitude, opt, Ti).

contrary to conventional hypotheses, that stage 1 may have little to do with slip processes. Indeed a further comment might be made in view of the evidence from plastic fatigue that when plastic strain produces slip it also produces misfit fissures in other directions, thus providing variously oriented fissures for an incipient crack to follow. This comment is that plastic strain is more likely to prevent a stage 1 shear crack than to establish it.

This conclusion is in keeping with behavior of what we are terming an elastic matrix, for it is one that in the absence of easy plastic flow lends itself to fracture by local buildups of internal stress and therefore to fracture processes in which slip need play no part. Figure 3 depicted such a process. There is also experimental support for this possibility of fracture by internal stress. This is shown by the next observation.

b. Internal Stresses. The above apparently slipless shear crack may leave a step in the boundaries of the grains that it traverses. Figure 32 shows an example. Apparently passage of the crack can relax a shear strain in each grain of a magnitude that might reasonably be taken as the length of the step divided by the width of the grain. It then turns out that this strain may reach values of the order 0.05, equivalent to the shear stress 0.05 G, a value that is

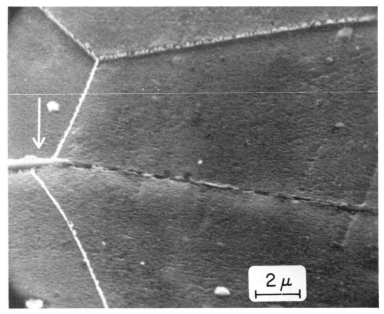

Fig. 32. Grain boundary shift by shear crack (alternating torsion, elastic amplitude, Ti).

not much less than the theoretical cohesive strength of metal, commonly put at 0.1 G (G is the shear modulus). A typical value for the elastic amplitude producing elastic fatigue in the titanium, from Figs. 10 and 11, is about 0.005 in terms of shear strain, or 0.005 G in terms of stress. So it appears that a relatively small amplitude is able to produce a local internal stress some ten times larger than itself, and large enough to rupture atomic bonds directly; a multiplying factor of 10 is a reasonable consequence of a mechanism like Fig. 3.

    c. A Tensile Stage 2. Further observations on the slipless shear crack show that when it becomes long enough it turns to a direction of maximum tension, then becoming a terminal crack. So the cracking in the elastic type of matrix obeys not only the stage 1 but also the stage 2 rule.

    The observations also suggest a physical reason for this turn to stage 2. First we recall that theories of crack propagation in general require that a crack tip should be the center of a stress concentration, and that this concentration by exceeding local yield should also produce around the crack tip a so-called plastic zone. Therefore cyclic stress should produce around the crack tip a zone of enhanced plastic or plasto-elastic fatigue. This in turn, by producing in the zone various local ruptures for the main crack-tip to link

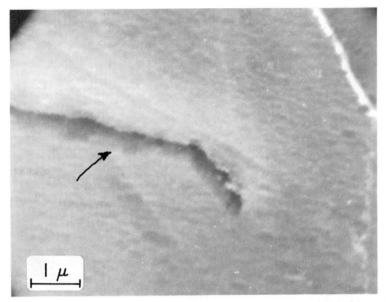

Fig. 33. Slipless shear crack entering elastic grain; no plastic zone ahead of tip (alternating torsion, elastic amplitude, Ti).

with, should aid its advance. Evidence for such a zone was shown in Fig. 29 for the case of plastic fatigue. The question arises whether a corresponding zone occurs in the case of elastic fatigue.

The answer seems to be that it need not do so while the propagating shear crack of stage 1 is small. For example, Fig. 33 shows the tip of such a shear crack that has halted part way through a slipless grain, where it is easy to see that the zone round its tip shows no sign of slip or plasticity, even at high magnification and after a light etch. Clearly the zone shows nothing comparable with Fig. 29, the case of plastic fatigue, and nothing in the way of enhanced cracking that might aid propagation.

It is possible, however, to observe signs of plasticity ahead of a small shear crack entering one of the occasional weak "plasto-elastic" grains that, as already noted, may coexist with the elastic ones. But the plastic effects still are not comparable with the case of plastic fatigue. All that appears is exemplified by Fig. 34, where a light etch has brought up points of enhanced reactivity here crowded between the crack tip and an adjacent grain boundary. These points may be significant when the crack passes through a plasto-elastic grain, but even these do not appear in an elastic grain. Therefore propagation of the stage 1 shear crack, at least in its early stages, must still depend on an elastic slipless mechanism such as the impulsive internal

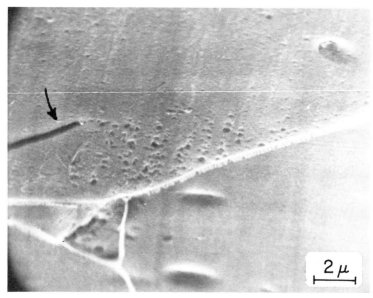

Fig. 34. Slipless shear crack entering plasto-elastic grain; points of fault concentration ahead of tip (alternating torsion, elastic amplitude, Ti).

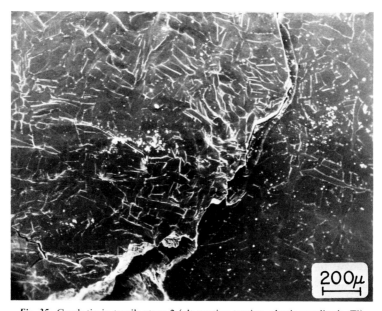

Fig. 35. Crack tip in tensile stage 2 (alternating torsion, elastic amplitude, Ti).

stress mechanism of Fig. 3 that starts it. Indeed that it should do so is reasonable. The internal stress can as readily propagate a crack as start one.

However, a plastic zone of sorts does appear when the shear crack of stage 1 becomes large enough to turn into the tensile direction of stage 2. This zone in the titanium is one that exhibits numerous grain-boundary fissures, now clearly the result of locally enhanced plasto-elastic fatigue. These are illustrated by Figs. 35 and 36. Figure 35 shows the tip of a sizeable crack in a field of grains whose boundaries, as shown by the higher magnification of

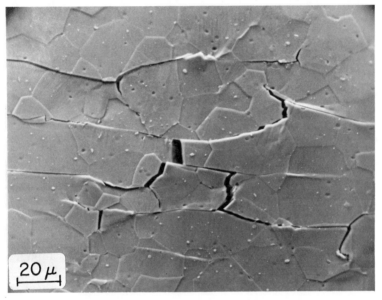

Fig. 36. Boundary cracking around crack tip of Fig. 35.

Fig.36, stand out because they are fissured; boundary fissures in fact were detectable up to about 1 mm from this crack tip, both ahead of the tip and all around it. It can be shown further, by again utilizing axial tension in the testing arrangement of Fig. 4, that a tension again pulls open especially those fissures at right angles to it. It then follows that propagation of the main crack is likely to be facilitated because the tip can link with fissure after fissure, and to be facilitated especially in a direction determined by maximum tension.

Whatever the final interpretation, it seems from the above observations that cracking in elastic fatigue differs from that in plastic fatigue not only in the way it propagates but also in the way it originates. In both types of fatigue, aside from accidents of structure, a crack must begin with local rupture of atomic bonds. Theoretical studies often emphasize the view that these bonds are too strong to be ruptured by any stresses that normal laboratory tests can apply. But this view assumes that the metal is homogeneous, when the applied stress must distribute itself over so many internal atomic bonds that it cannot significantly affect any single one. Hence, for rupture in practice, some inhomogeneity in some region must locally concentrate the stress. So far the inhomogeneity most favored by current theories has been a local confluence of dislocations that can coalesce and collapse and thus release energy enough not only for starting rupture but also for building the rupture up to a stable size (e.g., Cottrell, 1959). This mechanism could account readily for the plastic fatigue observed here. But it is less clear that it can account for the elastic fatigue, where rupture occurs in a matrix containing obstacles for the express purpose of suppressing any easy confluence of dislocations. The same obstacles, however, inevitably make the internal stresses inhomogeneous, at least while an external load is being applied. It is not too much to expect that an internal stress capable of local rupture is inevitable too. The mechanism of Fig. 3 is only one possibility. Thus the observations on elastic fatigue revive the question whether a rupture process based on dynamic internal stresses may not be more applicable to technical elastic metals than the conventional dislocation theories.

3. PLASTO-ELASTIC FATIGUE

Here the amplitude applied to elastic material exceeds yield and so should free many dislocations. But the obstacles responsible for technical elasticity still remain in the matrix to prevent the free and easy dislocation motion typifying simple plastic fatigue. Hence we may expect, at best, only imperfect attempts at the well-marked slip bands or slip-band fissures of plastic fatigue.

This inference is readily proved. If again to simplify comparisons we make observations on the weak grains exhibiting plasto-elastic fatigue in the above titanium under elastic amplitudes, we find that many of these grains during the whole fatigue life of a specimen develop no more than isolated point activities, the effect we have earlier termed hot spots. Figure 37 is typical, showing hot spots that are so isolated they hardly define the slip zones. Even when these point activities multiply and coalesce, they produce in most grains only short slip-band fissures of the kind illustrated after a light etch by Fig. 38. If this is plastic fatigue, it is plastic fatigue in very slow motion.

These inefficient plasto-elastic processes might ultimately lead to failure, given enough cycles and no competing mechanisms, but at elastic amplitudes they are superseded here by the more effective slipless cracking already

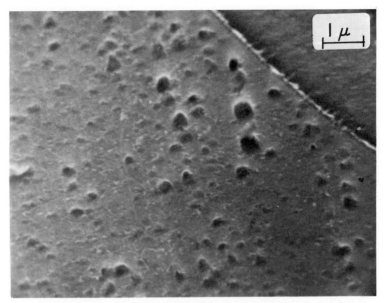

Fig. 37. Hot spots in plasto-elastic grain (alternating torsion, elastic amplitude, Ti, light etch).

Fig. 38. Slip-band cavities in plasto-elastic grain (alternating torsion, elastic amplitude, Ti, light etch).

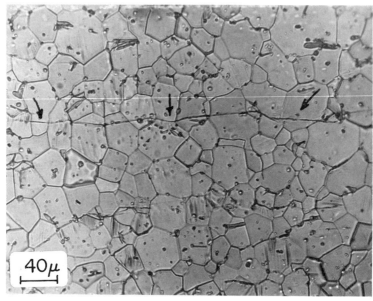

Fig. 39. Brittle transcrystalline shear crack in elastic-type metal under amplitude beyond yield (alternating torsion, Ti, opt).

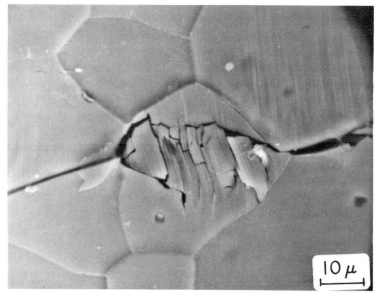

Fig. 40. Slip in plasto-elastic grain may interfere with traversing shear crack (alternating torsion, elastic amplitude, Ti).

described. But a further observation shows that the processes are still inefficient at amplitudes well in excess of yield. This is the observation illustrated by Fig. 39, where even at an amplitude twice as large as static yield the fracture process is still that of brittle shear. All the now more numerous plasto-elastic grains have done here is make it more difficult for a shear crack to pass through them. Slip-band fissures in the grains may interfere with the crack as in Fig. 40. Grain-boundary fissures, also appearing in the plasto-elasto range as in Fig. 36, may similarly interfere. But the observation that a brittle shear crack appears at all is itself enough to show that the elastic type of matrix is powerless to prevent high internal stresses even in its plastic range.

Titanium is a special case in that, having few slip systems, it is a metal in which plasticity in any event is hampered. It is possible that elastic matrices based on fcc and bcc structures and therefore having more slip systems might develop more effective slip-zone fissures during plasto-elastic fatigue, and fail as much from these fissures as from internal stresses and brittle shear. However, even in such structures it is not uncommon to find slip-band fissures and apparently brittle shear cracks developing independently in the same specimen. For example, Fig. 41 shows a slipless crack and Fig. 42 a

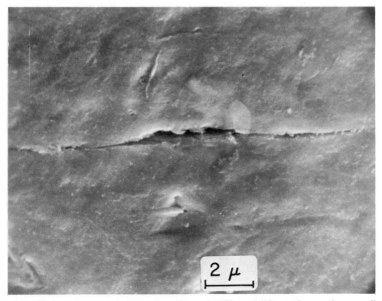

Fig. 41. Slipless shear crack in some regions of 1020 steel (alternating torsion, amplitude, near yield).

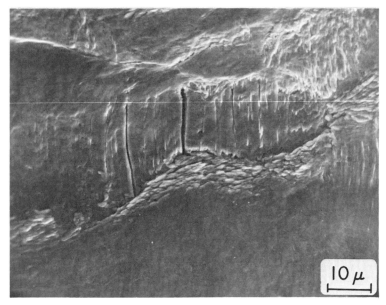

Fig. 42. Slip-band fissures in other regions of steel in Fig. 41.

slip-zone fissure in different regions of the same steel. Observations like this justify the classing of plasto-elastic fatigue as a process in its own right, possibly intermediate between pure elastic fatigue and pure plastic fatigue but definitely different from either.

4. ALLOY FATIGUE

Since every alloy is a special case, it might be misleading to concentrate on a particular one. But it is possible to notice that fatigue in most of the alloys studied in the literature seems to arise by one of two general processes. In both, the cyclic strain upsets the arrangement of alloying atoms in some region or zone. Then in one process the upset lowers the strength of that region; plasto-elastic fatigue sets in, and often an intense fissuring of slip bands comparable with that in plastic fatigue. Several of the aluminum alloys developed for aircraft exhibit fissures of this kind (e.g., Forsythe, 1969). In the other process, brittle shear cracks appear independently of any slip-band activities that may also occur, suggesting that here the crucial consequence of an unstable alloy composition is a local internal stress itself capable of rupture. Many of the complex steels exhibit such cracking (e.g., Ronay, 1971). In much alloy fatigue, too, the cracking again obeys the stage 1 and stage 2 rules. This is to be expected. One objective of alloying is to produce stable load-bearing materials. So their matrices necessarily are of the type we have described as elastic and therefore of a type that lends itself to developing internal stresses and brittle cracks.

C. Ultrasonic Fatigue

The fatigue so far discussed has been produced under simple standardized conditions. These were desirable; otherwise a proper comparison between the various types would be unnecessarily complicated. The next step in a fuller discussion would be to consider how the standard fatigue might be modified by systematically added variables. However, in practice, once the standard mechanism is clear, the modifications become largely predictable; examples were noted in Section II,B,1e. But one variable of technical as well as basic interest is a change in cycle frequency from the relatively low range used for most fatigue testing, generally less than some 150 Hz and often about 30 Hz, to the ultrasonic range of some 20,000 Hz or more.

One effect is that the fatigue mechanism may change. A basic factor in this change may be illustrated by plastic fatigue at the small F or S amplitudes. (We need consider only small amplitudes for the comparison, for these are all an ultrasonic apparatus can safely impose on a test specimen without overheating it.) First we recall that these small amplitudes in fatigue of annealed plastic metal at low frequency (LF) produce a mass of slip bands, slip-band fissures, and misfit fissures that make fracture a complex and long drawn-out process; illustrations appeared in Figs. 16–19, 28, and 29. The contrasting effect of comparable amplitudes at the high frequency (HF) is illustrated by Fig. 43. Here all that the HF amplitude has produced is an isolated slip-band. This restricted slipping is typical. It may consist of only one slip band in an area of many grains. But such an isolated band under ultrasonic cycles may rapidly fissure, propagate, and itself complete fracture before other slip bands and fissures can significantly multiply and interfere. So fracture by slip-band fissuring, an inefficient process at low frequency, becomes efficient at high frequency. Apparently the first HF fissure provides an almost exclusive sink for the HF energy (Mason and Wood, 1968).

The process tends to become somewhat more complex in the more plastic metals. The growing primary slip-band crack starts more secondary deformation in its neighborhood, and this, as for example in well-annealed copper, may slow down the crack and give time for slip bands to develop elsewhere. But the resulting slip even in highly plastic metal is still clearly different from the massive slip of normal LF plastic fatigue. The channeling of HF energy into a few preferred slip bands remains a distinctive and recognizable HF phenomenon.

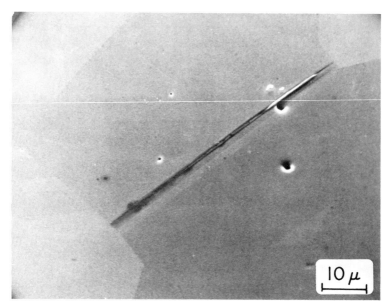

Fig. 43. Isolated slip-band starting crack in ultrasonic fatigue, in contrast with dense slip of low-frequency fatigue typified by Fig. 27 (70/30 brass).

Application of ultrasonic amplitudes to elastic metal produces a corresponding effect, with the surprising addition that the HF cycles can produce slip at lower amplitudes than LF cycles can. This appeared in work on low-carbon iron (Wood and Mason, 1969). We know that both HF and LF elastic amplitudes can free some dislocations from occasional pinning centers; some such freeing, as already discussed, is needed to account for LF elastic fatigue. At low frequency the dislocations thus freed normally are not numerous enough to produce the avalanche necessary for surface slip, though they may induce the local conditions for cracking by internal stress (Fig. 3). But at high frequency it appears that the dislocations first freed can act even at that stage as a sink for the HF energy, with the result that they become forerunners of an avalanche that starts local slip and the typical HF crack process. They do so even in the titanium where LF elastic amplitudes, as illustrated earlier, easily produce brittle shear cracks in slipless grains; in contrast, the HF fracture is always by slip, never by slipless shear (MacDonald and Wood, 1971).

The basic factor underlying these special HF effects seems to be that an HF cycle cuts down the time available for relaxation processes. For example, we noted when discussing slip-band fissuring in plastic fatigue that while some of the dislocations escaping from a source might reach the surface and start a fissure, others were likely to cross slip and form a strain-hardening atmosphere within the metal. But such cross slipping takes time, and this the HF cycle would cut down, thereby emphasizing the fissuring process at the expense of the hardening process and accounting for the efficient HF fissuring observed. Similarly we can account for why the HF fissure should propagate efficiently, for the HF cycle cuts down the time available for stresses at the fissure tip to relax by plastic flow, itself again essentially a time-dependent diffusion of lattice defects.

An informative experiment, suggested by this tendency of HF deformation to concentrate in preferred sinks, is to prestrain a metal in some way that should produce in it a large number of competing sinks. Subsequently applied HF cycles might then be unable to concentrate in any single one; the HF energy would be dispersed, and the HF fatigue inhibited. Tests of this kind, conducted on 70/30 brass (Gilmore et al., 1972), showed that the HF fatigue could in fact be inhibited, but only if the prestraining produced a uniform distribution of sinks. It was not sufficient, for example, merely to prestrain by cold drawing. This could prevent slip-band cracks, but it merely shifted the HF sinks and cracks to the grain boundaries. But inhibition could be achieved by prestraining that arranged the dislocations and lattice defects in a cell formation. In one procedure, the annealed brass was precycled at an H amplitude; in another, cold-drawn brass was precycled at an S amplitude small LF amplitudes being known to soften and rearrange a cold-worked dislocation structure. Then it was found that the HF deformation was harmlessly dispersed in a manner that, curiously enough, gave rise to a network of hot spots typically spaced at 10^{-4}–10^{-5} cm; Fig. 44 from cold-drawn brass that has been precycled at an S amplitude shows examples. These hot spots should mark the dispersed HF sinks, and it may be significant that their spacing is also that of the nodes in a typical dislocation cell structure.

Another interesting line of studies deals with whether ultrasonic frequencies might be utilized for the speeding up of fatigue tests now generally carried out at low frequency, for cycles that take many days to apply at low frequency take only a few minutes at the high one. At present the answer is not clear. At first sight the finding that LF and HF fatigue mechanisms may differ presents one difficulty, and the understandable consequence evident from several investigations that LF and HF s/n curves for the same material may also differ presents another (e.g., Kikukawa et al., 1965).

However, HF techniques may speed up tests on particular stages of fatigue. These, as already noted, are broadly three: an initiation stage in which many fissures may appear in many parts of a test specimen; an intermediate stage in which out of the fissures emerges a number of linkage cracks; and a final stage in which some linkage crack becomes large enough

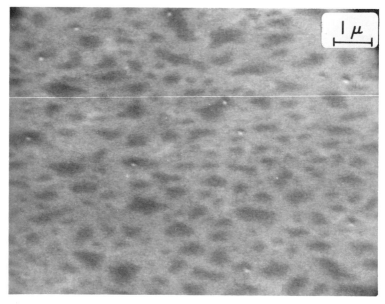

Fig. 44. Hot spots accompanying inhibited ultrasonic fatigue (cold-drawn 70/30 brass precycled at S amplitude).

to propagate itself and cause terminal fracture. The HF test may prove useful for study of the initiation stage and for comparing how quickly it may appear in different materials, or in different samples of the same material, for its appearance is indicated sensitively by increase in HF energy absorption or internal friction Q^{-1}. It may be reasonable to expect that if HF fatigue starts easily in a particular material then LF fatigue is likely to do so too. But whether HF tests can be useful for the following two stages is another matter. Crack growth is time dependent, depending in particular on atmospheric attack, an effect that alone would make HF and LF s/n curves very different. It would seem that for both the intermediate and propagation stages the design engineer has little option but to rely still on data produced under conditions that most closely simulate his requirements.

III. Concluding Survey

As noted below, many excellent reviews of basic fatigue already exist, but few of them make clear the types of fatigue with which they are concerned. Actually it seems inevitable from the way study of fatigue began and then ramified that distinctions among its different forms should have become blurred. Thus systematic studies began in the middle of the last century when the speed age began, and engineers noticed that moving parts like axles and wheels could fail as much from stress repetition as from stress magnitude. But the main structural material for these parts was plain steel, or what here we should class as the elastic type of metal, so that the principles revealed by these early studies were primarily those underlying elastic fatigue only. These first studies lasted until the early 1900s, and the concepts arising from them are embodied in the contemporary book by Gough (1926).

Then the emphasis shifted to quite a different material as large single crystals of metal became available, for the crystals most easily grown were those of what here we have described as plastic metals, for example copper and aluminum, and so these next studies were primarily of plastic fatigue. The part of surface slip was emphasized; fatigue was said to occur only where slip concentrated; and it occurred as a result of local hardening by the slip. These and other concepts of the day appear in a well-known review by Gough (1933). At the same time observations that did not agree too well with the concepts appeared; the slip bands in iron, for example, might not produce well-defined fissures; but before these difficulties could be resolved another shift occurred to a different material again.

This was the shift to fatigue of the light alloys then being developed for the growing aircraft industry and, since fatigue in most of these alloys arose partly from local chemical instabilities, a shift in effect to what here we have termed alloy fatigue. The instabilities, and a common susceptibility of these alloys to corrosion fatigue, introduced entirely new concepts, and they have also made it more difficult to conceive any unified theory of fatigue. Though some common principles are recognizable, many usefully discussed by Forsythe (1969), the literature on the alloys is essentially one of special cases.

Thus by the 1950s a mass of information on fatigue phenomena had accumulated. Perhaps the most notable attempt to reduce the data to some kind of order was the valuable review by Thompson and Wadsworth (1958). But many complexities remained; the part played by slip bands was still overemphasized, and the possibility of a virtually slipless fatigue would not have been admitted. And about this time too the ways of interpreting slip, fatigue, and deformation generally had altered, partly because of developing theories about how deformation depended on dislocations and other lattice defects, but mainly because of developing techniques, particularly the thin foil transmission technique of electron microscopy, that made it possible to see the lattice defects directly. This altered viewpoint underlies useful reviews by Ham (1966) and by Segall (1968). The direct observations by electron transmission may not have lived up to their early promise; the thin films are too sensitive to processing, and defect distributions shown by films extracted from various parts of a specimen differ too much (indicating incidentally that what matters more in fatigue than the defects in the very small regions that the transmission technique concentrates on is the statistical distribution of defects in much larger regions, just as what matters more in determining the pressure of a gas than molecular velocities in one very small volume is the statistical mean velocity in a large volume). But observations have shown which distributions are probable and significant and have established dislocation theory as the only one capable of sometime providing a unified treatment of fatigue effects.

The direction basic fatigue studies may take in the future is likely to be influenced by the need to understand effects of more technical interest than the simple effects naturally studied first. Examples are those effects underlying crack propagation in the later stages that in recent years have mainly occupied the attention of design engineers (1970, 1972); and especially those effects produced by fluctuating stress amplitudes or fluctuating test temperature or other fluctuating conditions that can cause a metal microstructure and its defect distribution to change from one relatively stable state to another, for various observations indicate that a metal during the change seems unusually weak and abnormally susceptible to external stress. An extreme illustration is the superplasticity effect, or the lowering of creep resistance during recrystallization or recovery; but, though less extreme, a similar weakening can appear during the above-mentioned kind of fluctuating test conditions (Wood and Utili, 1969). In the past what has been studied most in basic fatigue research is the effect, for example, of a stress amplitude that is constant. Sometimes a test specimen has been precycled at one amplitude and then subjected to another. But what still has been omitted is study of the effects occurring only while the amplitude is changing from one value to the other and the microstructure is in a state of flux.

References

Cottrell, A. H. (1959). In "Fracture" (B. L. Averback *et al.*, eds.), p. 20. Wiley, New York.

Forsythe, P. J. E. (1962). *Symp. Crack Propagation, Coll. Aeronaut., Cranfield, 1962,* **1**, 76.

Forsythe, P. J. E. (1969). "Physical Basis of Metal Fatigue." Blackie, Glasgow and London.

Gilmore, C. E., MacDonald, D. E., and Wood, W. A. (1972). *Acta Met.* **20**, 953.

Gough, H. J. (1926). "Fatigue of Metals." Benn, London.

Gough, H. J. (1933). ASTM, "Marburg Lecture."

Grosskreutz, J. C., Reimann, W. H., and Wood, W. A. (1966). *Acta Met.* **14**, 1549.

Ham, R. K. (1966). *Can. Met. Quart.* **5**, 161.

Kikukawa, M., Ohji, K., and Ogura, K. (1965). *J. Basic Eng.* **87**, 857.

Liebowitz, H., ed. (1969–1971). "Fracture," Vols. 1–7. Academic Press, New York.

MacDonald, D. E., and Wood, W. A. (1971). *J. Appl. Phys.* **42**, 5531.

MacDonald, D. E., and Wood, W. A. (1972). *J. Inst. Metals* **100**, 73.

Mason, W. P., and Wood, W. A. (1968). *J. Appl. Phys.* **39**, 5581.

Ronay, M. (1971). *In* "Fracture" (H. Liebowitz, ed.), Vol. 3, p. 431. Academic Press, New York.
Segall, R. L. (1968). *Advan. Mater. Res.* **3**, 109.
Symp. Fracture Fatigue (1972). *George Washington Univ., Washington, D.C.*
Thompson, N., and Wadsworth, N. J. (1958). *Advan. Phys.* **7**, 72.
Wood, W. A. (1971). "Study of Metal Structures and their Mechanical Properties." Pergamon, Oxford.
Wood, W. A., and Mason, W. P. (1969). *J. Appl. Phys.* **40**, 4514.
Wood, W. A., and Segall, R. L. (1957). *Proc. Roy. Soc. (London) Symp. Internal Stresses.*
Wood, W. A., and Utili, F. (1969). *Int. Conf. Weld. Metal Test., Roumania* **2**, 305.

The Relationship between Atomic Order and the Mechanical Properties of Alloys

M. J. MARCINKOWSKI

Engineering Materials Group, and
Department of Mechanical Engineering
University of Maryland
College Park, Maryland

I. Introduction

The first demonstration of the effect of atomic ordering on mechanical behavior was carried out with Cu_3Au single crystals by Sachs and Weerts[1] in 1931 at The Kaiser Wilhelm Institut für Metallforschung in Berlin. Eight years earlier, in 1923 Bain[2] first showed experimentally by means of X-ray techniques that long-range atomic order existed in Cu_3Au, or in any alloy for that matter. These investigations in turn were motivated by the work of Tammann[3] at Göttingen who first postulated the existence of long-range atomic order as early as 1919. The development of dislocation theory during the 1950s and the introduction of transmission electron microscopy during the 1960s provided further motivation for the study of the relationship between atomic order and mechanical properties.

A number of reviews[4–8] and symposiums[9,10] dealing with atomic order have appeared over the years. Most of the earlier reviews[4–6] have devoted only a small fraction of their space to the mention of mechanical properties. One of these reviews[5], in fact, had gone so far as to state that the process of ordering does not in itself seem to affect the gross mechanical properties of alloys.

It will be shown in the following sections that atomic ordering can lead to profound changes in the plastic deformation and fracture behavior of alloys. Furthermore, these effects will be treated from an atomistic or microscopic point of view utilizing fully the concepts developed within the framework of dislocation theory.

II. The Nature of Dislocations in Ordered Alloys

A. Straight Dislocations

Since an extensive review of dislocation morphologies in ordered alloys has already been carried out[11], it will suffice here to underscore only some of the more important points. In the first place, a dislocation that is perfect in a disordered alloy, i.e., possesses a Burgers vector that makes the crystal lattice invariant under the translation, will not necessarily be perfect when the alloy becomes ordered. This follows from the fact that atomic ordering alters the periodicity of the original disordered unit cell. The dislocation that was perfect in the disordered alloy now becomes imperfect in the corresponding ordered alloy and must therefore be attached to a planar fault, i.e., the so-called antiphase boundary (APB). In order to minimize the high-energy APB, it may terminate on a second imperfect dislocation. The resulting finite length of the APB bounded by a pair of imperfect dislocations is termed a superlattice dislocation and is illustrated schematically in Fig. 1a. Depending upon the nature of the crystal lattice and the type of atomic ordering, the superlattice dislocation may be more complex as shown in Fig. 1b, where it is seen to consist of four imperfect dislocations coupled by two different types of APB as indicated by different type cross hatchings.

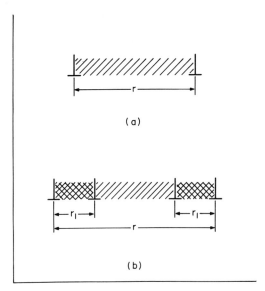

Fig. 1. Simple (a) and more complex (b) type of superlattice dislocations in ordered alloys.

Superlattice dislocations were first predicted theoretically in 1947 by Koehler and Seitz[12] and were first observed experimentally in ordered Cu_3Au by means of transmission electron microscopy (TEM) in 1960[13–15].

The equilibrium configuration of a superlattice dislocation can be obtained by first writing down its total energy E_T. A calculation of this type was first carried out by Brown and Herman[16] in 1956. In particular,

$$E_T = \sum E_S + \sum E_I + \sum E_{\gamma_0} \tag{1}$$

where the first summation represents the self-energy of the individual dislocations comprising the superlattice dislocation, the second summation is the interaction energy between all of these dislocations, and $\sum E_{\gamma_0}$ is the sum of all the APB energies associated with the superlattice dislocation. For straight screw dislocations in an isotropic medium[17],

$$E_{S,S} = \frac{\mu b^2}{4\pi} \ln \frac{\alpha l}{b} \tag{2a}$$

whereas for straight edge dislocations,

$$E_{S,E} = \frac{\mu b^2}{4\pi(1 - v)} \ln \frac{\alpha l}{b} \tag{2b}$$

where μ is the isotropic shear modulus, b the Burgers vector of an individual imperfect dislocation, v is Poisson's ratio, l is the shortest distance from the dislocation to the free surface, and α is a dislocation core parameter. All energies are expressed in terms of a unit length of dislocation. The interaction energy between two parallel straight dislocations separated by a distance r is given by[17]

$$E_1 = -\frac{\mu(\mathbf{b}_1 \cdot \xi)(\mathbf{b}_2 \cdot \xi)}{2\pi} \ln\frac{r}{l} - \frac{\mu}{2\pi(1-v)}[(\mathbf{b}_1 \times \xi) \cdot (\mathbf{b}_2 \times \xi)] \ln\frac{r}{l}$$

$$- \frac{\mu}{2\pi(1-v)r^2}[(\mathbf{b}_1 \times \xi) \cdot \mathbf{r}][(\mathbf{b}_2 \times \xi) \cdot \mathbf{r}] \tag{3}$$

where $\mathbf{b}_1$ and $\mathbf{b}_2$ refer to the Burgers vectors of the two interacting dislocations and ξ is the line direction associated with the dislocations. For a pure screw dislocation pair of like Burgers vector, Eq. (3) reduces to

$$E_{1,s} = \frac{\mu b^2}{2\pi} \ln\frac{l}{r} \tag{4a}$$

whereas for a pure edge dislocation pair of like sign

$$E_{1,E} = \frac{\mu b^2}{2\pi(1-v)}\left[\ln\frac{l}{r} - \sin^2\theta\right] \tag{4b}$$

where θ is the angle that $\mathbf{r}$ makes with the direction of $\mathbf{b}$. For the case of the relatively simple superlattice dislocation illustrated in Fig. 1a which is extended on the glide plane, $\theta = 0°$ in Eq. (4b), and Eq. (1) becomes

$$E_{T,M} = \frac{\mu b^2 L}{2\pi}\left[\ln\frac{\alpha l}{b} + \ln\frac{l}{r}\right] + \gamma_0 r \tag{5}$$

where L is an orientation factor given by

$$L = \cos^2\theta + \frac{\sin^2\phi}{1-v} \tag{6}$$

where ϕ is the angle that the Burgers vector makes with the line direction vector ξ, so that for screw dislocations $\phi = 0$, while for edges it is 90°. The equilibrium extension of the superlattice dislocation shown in Fig. 1a is obtained from the condition $\partial E_T/\partial r = 0$, which when applied to Eq. (5) gives

$$r_e = \mu b^2 L/2\pi\gamma_0 \tag{7}$$

This is a stable equilibrium since $\partial^2 E_T/\partial r^2 > 0$, i.e. the energy possesses a minimum. In the case of the somewhat more complex dislocation shown in Fig. 1b the equilibrium configuration is a function of two variables r and r_1 so that the conditions for equilibrium are given by the pair of equations $\partial E_T/\partial r = 0$ and $\partial E_T/\partial r_1 = 0$ which must be satisfied simultaneously. In such cases the equilibrium values of r and r_1 can be obtained by numerical, i.e. computer, techniques[18]. The APB energy γ_0 can in general be written as

$$\gamma_0 = Mv \tag{8}$$

where M is a geometrical factor determined by the particular crystallographic plane hkl on which the APB lies as well as the crystal structure, while v is given by

$$v = v_{AB} - \tfrac{1}{2}(v_{AA} + v_{BB}) \tag{9}$$

where v_{AB}, etc., are the energies of a nearest neighbor AB bond, etc. An analytic expression for v can be written as

$$v = kT_C/N \tag{10}$$

where k is Boltzmann's constant, T_C is the critical ordering temperature, and N is a parameter which depends upon both the crystal structure as well as the particular theory. Various values of N have been tabulated by Guttman[19], whereas various expressions for M (hkl) have been listed in an earlier review article by the present author[20].

Thus far, consideration has been given to dislocations in isotropic materials. In order to introduce anisotropy into the elastic energy contributions, it is necessary only to replace μ in Eqs. (2a) and (4a) by K_S and $\mu/(1 - v)$ in Eqs. (2b) and (4b) by K_e, where K_S and K_e are the energy factors for pure screw and edge dislocations, respectively[17]. Thus, for an anisotropic crystal, Eqs. (5) and (7) can be rewritten as

$$E_{T, M, A} = \frac{b^2 K}{2\pi} \left[\ln \frac{\alpha l}{b} + \ln \frac{l}{r} \right] + \gamma_0 r \tag{11}$$

and

$$r_{e, A} = b^2 K / 2\pi\gamma_0 \tag{12}$$

where K is defined as

$$K = K_S \cos \phi + K_e \sin \phi \tag{13}$$

Combining Eqs. (12), (8), and (10), we obtain

$$r_{e, A} = \frac{b^2 KN}{2\pi M K T_C} \tag{14}$$

It follows from this relation that for T_C sufficiently high $r_{e, A}$ reduces to the point where the two dislocations can be considered as combined into a single dislocation. Under these conditions, $E_{T, M, A}$ given by Eq. (11) may be sufficiently large so that another type of dislocation with different Burgers vector may be energetically more favorable. Arguments of this type were first used by Rachinger and Cottrell[21,22] to postulate that for small ordering forces, i.e. low values of T_C, dislocations in CsCl type superlattices consisted of superlattice dislocations such as that shown in Fig. 1a where each of the ordinary dislocations had Burgers vectors of the type $\tfrac{1}{2}a_0\langle 111 \rangle$ where a_0 is the unit cell edge length of the crystal. On the other hand, for large ordering forces, i.e. large values of T_C, single perfect dislocations of the type $a_0\langle100\rangle$ are more stable. The relative stability between various dislocation configurations can be determined only by comparing their total energies as given by expressions of the type shown in Eq. (1). Attempts to predict the slip system based upon relationships of the type given by Eq. (1) have been made by a number of investigators[23–26]. A serious omission in Eq. (1) is an energy term E_τ due to the externally applied stress τ. This is a rather complex contribution and depends upon the details of the particular dislocation configuration. As an example of two extremes, consider just the case where τ is applied parallel to the slip plane of the superlattice dislocation shown in Fig. 1a. Since the applied force τb due to this stress remains the same on each dislocation due to the fact that the Burgers vectors of the two dislocations are parallel, the equilibrium configuration of the superlattice dislocation is unaffected by the applied stress. If, on the other hand, single crystals of NiAl[27] and AuZn[28], which normally glide along $\langle001\rangle$, are oriented with their axes along [001], slip occurs along $\langle111\rangle$. The applied stress will be examined more explicitly in some of the analyses to follow.

Substitution of Eq. (12) into the last term of Eq. (11) shows that the APB energy term is small compared to the other two terms, i.e., the self and interaction energies. On the other hand, the importance of the APB energy term shows up in its influence on the interaction energy, i.e., the second term in Eq. (11).

Finally, it would appear that the crystal anisotropy might be of significance in affecting the morphology of the dislocations in an ordered alloy. Figure 2 shows the energy of three dislocations lying on the $(\bar{1}10)$ slip plane in β-brass with the Burgers vectors indicated, i.e., $a_0[111]$, $a_0[110]$, and $a_0[001][29]$. Thus, $r_{e,A}$ given by Eq. (14) is assumed to be zero in these calculations; i.e., the superlattice dislocation is not extended as in Fig. 1a. The angle θ designates the orientation of these dislocations measured with respect to the [001] direction, while the unit of energy associated with the ordinate is given as

$$E_{S,S}^{[001]} = \frac{a_0^2 C_{44}}{4\pi}\left(\frac{\alpha l}{b}\right) \tag{15}$$

where C_{44} is one of the three elastic stiffness constants associated with a cubic crystal. The anisotropy factor A, given by[30]

$$A = \frac{2C_{44}}{C_{11} - C_{12}} \tag{16}$$

is 8.75 for β-brass, a relatively high value. It would appear, according to Fig. 2, that depending upon θ, any of the three dislocations indicated therein

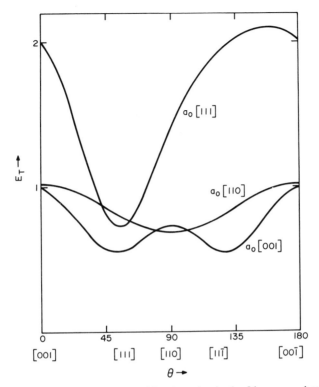

Fig. 2. Variation of dislocation energy with orientation in the β-brass superlattice for the three Burgers vectors shown. From Head[29].

could be stable relative to the others. On the other hand, if the $a_0[111]$ dislocation were allowed to dissociate into two $\frac{1}{2}a_0[111]$ dislocations according to Fig. 1a, as in fact it does[31], then the entire [111] curve shown in Fig. 2 would drop due to the reduction in elastic energy. The superlattice screw dislocation would then be expected to possess minimum energy.

In the case of NiAl, which also possesses a β-brass type superlattice, a set of curves similar to those shown in Fig. 2 can be generated[32]; these are illustrated in Fig. 3. Because of the relatively low value of A (3.28 for NiAl), the variations in $E_{s,s}^{[001]}$ are not as pronounced as was the case for β-brass. In addition, the [001] dislocation always exhibits the lowest energy of the three dislocations shown. Furthermore, because of the very high value of T_C for this alloy (T_C is taken as its melting point, 1913°K), $r_{e,A}$ given by Eq. (14) is expected to be zero, which is the basis upon which the curves in Fig. 3 were calculated. Thus, the [111] curve in Fig. 3 cannot be lowered significantly by dislocation dissociation and so unlike the case of β-brass where the slip direction is $\langle 111 \rangle$[31], it is $\langle 100 \rangle$ in the NiAl superlattice[32]. These results

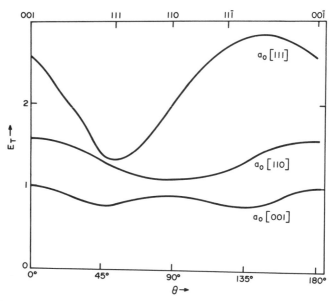

Fig. 3. Variation of dislocation energy with orientation in the NiAl superlattice for the three Burgers vectors shown. From Lloyd and Loretto[*32*].

can be seen more clearly in Table I where the designations NN, NNN, and SF in the last column signify nearest neighbor, next nearest neighbor, and stacking fault, respectively[*11,20*]. Table II, on the other hand, lists the various parameters relevant to an understanding of the dislocation morphologies associated with a given ordered alloy.

The crystalline anisotropy also gives rise to an instability in the dislocations for certain values of θ, e.g., $\theta = 0$ for [001] dislocations in NiAl[*32,33*]. The resulting irregularity in the dislocation line could give rise to a modification in the energy terms of Eq. (1). However, for simplicity this effect will be neglected. In summary, then, it is seen thus far by reference to Eq. (11) that b, K, T_C, and τ all strongly affect the morphology of dislocations in ordered alloys. Tables I and II illustrate a variety of these dislocation morphologies associated with a number of different types of ordered alloys along with the relevant physical parameters.

One of the most profound effects of the strong variation of E_T with θ for the [111] dislocation in Figs. 2 and 3 is that the energy for an edge dislocation is more than twice that for a screw type dislocation. This means that there should be a preference for screw dislocations over edge dislocations in crystals where the Burgers vectors lie along [111]. That this is indeed so was verified in the TEM studies carried out with Fe_3Si[*34*] and Fe_3Al[*36*]. A

TABLE I

DISLOCATION CONFIGURATIONS IN VARIOUS ORDER ALLOYS

Superlattice type (*Strukturbericht* designation)	Chemical designation	Unit cell dimensions	Alloy types	Superlattice dislocation type	Burgers vector of each dislocation	Antiphase boundary type
B2	CsCl	a_0	NiAl, AgMg AuZn	⊥	$a_0\langle 100\rangle$	None
			CuZn, FeCo FeAl, FeRh NiAl, AgMg AuZn	⊥ …… ⊥	$\frac{1}{2}a_0\langle 111\rangle$	NN
DO₃	Fe₃Al	a_0	Fe₃Al, Fe₃Si Fe₃B	⊥ ▨ ⊥ …… ⊥	$\frac{1}{4}a_0\langle 111\rangle$	NN; NNN
L1₂	Cu₃Au	a_0	Cu₃Au, Ni₃Mn Ni₃Al, Ni₃Fe Cu₃Pd, Ni₃Ti Ag₃Mg, Ni₃Ta Ni₃Si, Cu₃Pt Ni₃Ga	▨ ⊥ …… ⊥ ▨	$\frac{1}{6}a_0\langle 112\rangle$	NN; NN + SF
DO₁₉	Mg₃Cd	a_0 c_0	Mg₃Cd	…… ⊥ ▨ ⊥; …… ⊥	$\frac{1}{6}a_0\langle 10\bar{1}0\rangle$; $\frac{1}{2}a_0\langle 2\bar{1}\bar{1}0\rangle$	NN; SF; NNN
L1₀	CuAu	a_0 c_0	CuAu, CoPt FePt	▨ …… ⊥ ▨ ……; ▨ ⊥	$\frac{1}{6}a_0\langle 112\rangle$	NN; NN + SF; SF

TABLE II

PARAMETERS ASSOCIATED WITH DISLOCATIONS IN ORDERED ALLOYS

Superlattice type (Strukturbericht designation)	Alloy types	$T_C(°K)$	$T_M(°K)$	$\dfrac{T_C}{T_M}$	($\times 10^{12}$ dyn/cm²)			A	C'	(ergs/cm²)	
					C_{11}	C_{12}	C_{44}			γ_{NN}	γ_{NNN}
B2	NiAl	1913	1913	1.00	2.115	1.432	1.121	3.28	0.342	NO APB	
	AgMg	1093	1093	1.00						NO APB	
	AuZn	998	998	1.00						NO APB	
	CuZn	738	1148	0.643	1.19	1.02	0.745	8.75	0.085	40[a]	
	FeCo	1003	1708	0.59						157[b]	
	FeAl	1573	1573	1.00						>105[c]	
	FeRh	1573								High	
	NiTi		1583							High	
	NiAl	1913	1913	1.00	2.115	1.432	1.121	3.28	0.342	High	
	AgMg	1093	1093	1.00						High	
	AuZn	998	998	1.00						High	
DO₃	Fe₃Al	848	1753	0.474	1.7099	1.3061	1.3170	6.53	0.2019	77[d]	85[d]
	Fe₃Si	1503	1503	1.00	2.3409	1.5627	1.3555	3.49	0.3891		225[e]
	Fe₃B	620	1573	0.394							

L1$_2$	Cu$_3$Au	661	1223	0.54	2.25	1.73	0.663	2.55	0.26	91[f]
	Ni$_3$Mn	753	1498	0.504						123[g]
	Ni$_3$Al	1673	1673	1.00						300[h]
	Ni$_3$Fe	773	1708	0.452						
	Cu$_3$Pd	743	1393	0.533						
	Ag$_3$Mg	660	1073	0.615						
	Ni$_3$Ta	1623	1818	0.895						
	Ni$_3$Si	1503	1503	1.00						
	Cu$_3$Pt	873	1553	0.561						40[i]
	Ni$_3$Ga	1523	1523	1.00						
DO$_{19}$	Mg$_3$Cd	426	773	0.552						
L1$_0$	CuAu	683	1163	0.586						
	CoPt	1098	1773	0.620						
	FePt	1573	1853	0.850						

[a] Head et al.[31].
[b] Marcinkowski and Chessin[55].
[c] Ray et al.[97].
[d] Crawford et al.[96].
[e] Lakso and Marcinkowski[34].
[f] Marcinkowski et al.[15].
[g] Marcinkowski and Miller[58].
[h] Taunt[98].
[i] Takeuchi and Kuramoto[151].

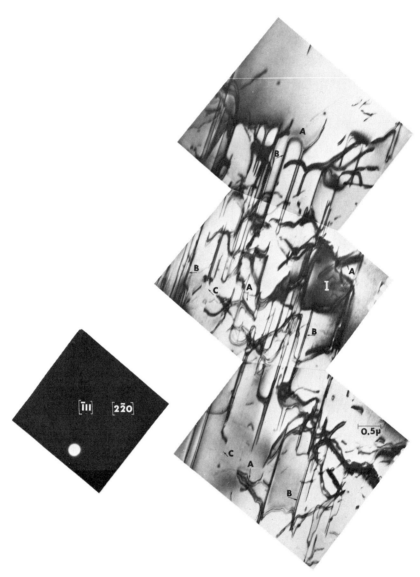

Fig. 4. Bright field micrograph of superlattice dislocations in Fe$_3$Si after 2.2% strain at 298°K showing both screw (B) and edge (A) dislocation locks as well as annihilation of screw segments (C); normal to foil is [110]. From Lakso and Marcinkowski[*34*].

typical dislocation configuration lying in the [110] glide plane is shown in Fig.4 for a slightly deformed Fe_3Si superlattice. The screw segments (labeled B) are seen to be highly elongated, while the edge segments (labeled A) are correspondingly shorter. Many of the screw segments are seen to have cross-slipped out of the plane of the foil (C). The source for many of the dislocations in Fig. 4 is believed to be the inclusion labeled as I.

It has been argued that the energy, as given by Eq. (1), is alone insufficient to predict the slip system associated with a given superlattice and that the dislocation mobility must also be considered[23]. The mobility is defined as the ratio of the stress required to move the dislocation to the stress required to shear the slip plane rigidly. It is given by[35]

$$S = 4\pi(\zeta/b) \exp\left[-2\pi(\zeta/b)\right] \tag{17a}$$

where ζ is the width of the dislocation and can be expressed as[36]

$$\zeta = dK/2C \tag{17b}$$

where d is the spacing between the glide planes and C is the shear modulus in the slip direction on the glide plane. Equation (17a) is in effect related to the Peierls (lattice friction) stress required to move a dislocation in an otherwise perfect crystal. In order that it be handled properly, it must be included in the expression for the total energy of Eq. (1), and is in fact one aspect of the stress term discussed earlier. It is extremely difficult, however, to calculate this Peierls stress from first principles[37]. It is interesting to note that in the case of β-brass[23] and Fe_3Al[38], where the slip direction is along $\langle 111 \rangle$, S for edge dislocations is greater than that of screws and would thus give dislocation morphologies just the reverse of that shown in Fig. 4, i.e., elongated edge segments and shortened screw segments. However, as will be shown later in this review, the Peierls stress may not be properly represented by a relation of the type given by Eq. (16). Instead, it may be more closely related to the manner in which the dislocation can dissociate on various cross slip planes.

B. Dislocation Loops

In the previous sections straight dislocations were considered, so that the line tension of the dislocation was constant. When a dislocation loop is considered, this is not the case, and the line tension of the dislocation becomes a function of its size. This leads to significant alterations in the previous calculations. For simplicity, only circular dislocation loops will be considered, and the circular analogue of the straight superlattice dislocation shown in Fig. 1a is illustrated in Fig. 5a. The present treatment of circular

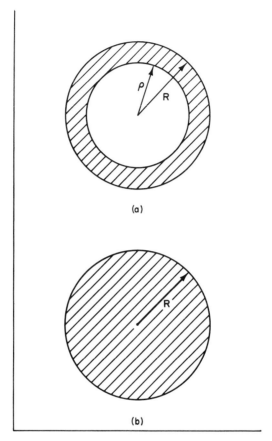

Fig. 5. (a) Circular counterpart of the straight superlattice dislocation shown in Fig. 1a. (b) Ordinary dislocation loop in an ordered alloy.

dislocations in superlattices[39] is based upon the methods of Kröner[40–42]. In particular, the self-energy of a circular dislocation loop can be written as

$$E_{s, c} = \frac{2m - 1}{2(m - 1)} \mu b^2 \frac{R}{2} \left[\ln \left(\frac{4R}{\varepsilon} \right) - 2 \right] \tag{18}$$

where $m = 1/v$, R is the radius of the loop, and ε is a dislocation core cut-off parameter given approximately by $2b$. The term $(2m - 1)/2(m - 1)$ in Eq. (18) is an orientation factor which accounts for the equal edge and screw components of the dislocation loop.

The interaction energy between two coaxial dislocation glide loops of radius R and ρ lying on the same slip plane is given by

$$E_{I,C} = \frac{2m-1}{2(m-1)} \mu b^2 (R+\rho) \left[\left(1 - \frac{k^2}{2}\right) \tilde{K} - \tilde{E} \right] \tag{19}$$

where $\tilde{K}$ and $\tilde{E}$ are complete elliptic integrals of the first and second kind, respectively. The modulus k is given by

$$k^2 = 4\rho R/(R+\rho)^2 = 1 - k'^2 \tag{20}$$

where k' is the complementary modulus. When $R \to \infty$ or when $(R - \rho) \to 0$, $k \simeq 1$ so that $\tilde{E} \simeq 1$ and $\tilde{K} \simeq \ln(4/k')$. In addition, $\rho \simeq R$. Substituting these approximations in Eq. (19) gives

$$E_{I,C,R\to\infty} = \frac{2m-1}{2(m-1)} \mu b^2 R \left[\ln\left(\frac{8R}{r}\right) - 2 \right] \tag{21}$$

where r is the spacing between the two ordinary dislocations that comprise the superlattice dislocation loops in Fig. 5a, i.e., $r = R - \rho$. The total energy associated with the superlattice dislocation loop in Fig. 5a is given by

$$E_{T,C} = \sum E_{S,C} + E_{I,C} + \pi \gamma_0 (R^2 - \rho^2) - \pi \tau b (R^2 + \rho^2) \tag{22}$$

The first three terms in the above equation are analogous to the first three terms in Eq. (1). However, in order to maintain the dislocation loops in equilibrium, it is necessary to add the fourth term which takes into account the work done on the system by the applied stress. This was not necessary for the straight dislocation illustrated in Fig. 1a. The condition for equilibrium in Eq. (22) is that $\partial E_{T,C}/\partial R = 0$. Using the approximations $(R^2 - \rho^2) \simeq R r_e$ and $(R^2 + \rho^2) \simeq 2R^2$ we obtain

$$\tau_{R\to\infty} = \frac{2m-1}{2(m-1)} \frac{\mu b}{4\pi R} \left[\frac{1}{2} \ln\left(\frac{4R}{\varepsilon}\right) + \ln\left(\frac{8R}{r_e}\right) - \frac{3}{2} \right] + \frac{\gamma_0 r_e}{2Rb} \tag{23}$$

The above equation gives the stress required to hold the superlattice dislocation loop of radius R shown in Fig. 5a in equilibrium. It is an unstable equilibrium since $\partial E_{T,C}^2/\partial R^2 < 0$, i.e., there is a maximum in $E_{T,C}$.

A second important type of dislocation loop that may exist within a superlattice is the imperfect loop shown in Fig. 5b which bounds on APB. The total energy of this loop is somewhat simpler than that given by Eq. (22). In particular:

$$E_{T,C} = E_{S,C} + \pi R^2 (\gamma_0 - \tau b) \tag{24}$$

applying the conditions for equilibrium, i.e., $\partial E_{T,C}/\partial R = 0$ gives

$$\tau = \frac{2m - 1}{2(m - 1)} \frac{\mu b}{4\pi R} \left[\ln \left(\frac{4R}{\varepsilon} \right) - 1 \right] + \frac{\gamma_0}{b} \tag{25}$$

Equations similar to those given by (23) and (25) have also been developed by Ashby[43]. Figure 6 shows a typical $E_{T,C}$ versus R curve for a single dislocation in a disordered alloy, i.e., $\gamma_0 = 0$ in Eq. (24) under an applied stress $\tau = 100 \times 10^8$ dyn/cm². The maximum in the curve corresponds to the condition of unstable equilibrium and gives the dislocation loop size R_C and energy E_T associated with this equilibrium at the given stress level of

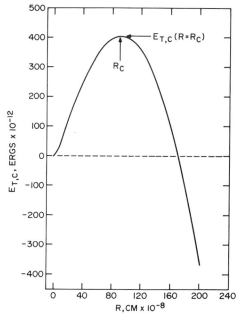

Fig. 6. Typical energy versus radius plot for an ordinary dislocation in a disordered alloy under an applied stress of $\tau = 100 \times 10^8$ dyn/cm². From Marcinkowski and Leamy[39].

$\tau = 100 \times 10^8$ dyn/cm², which could have also been obtained from equation 25 by letting $\gamma_0 = 0$. From curves of the type shown in Fig. 6, plots of R_C versus τ_{RC} were obtained and are shown for three dislocation loop configurations as indicated in Fig. 7. The numerical values of the constants used in the calculations are given in Marcinkowski and Leamy[39]. It is apparent from the curves that atomic ordering increases the stress level required to maintain a dislocation loop at a fixed size. Furthermore, above stress levels of about 150×10^8 dyn/cm², the superlattice dislocation loop is

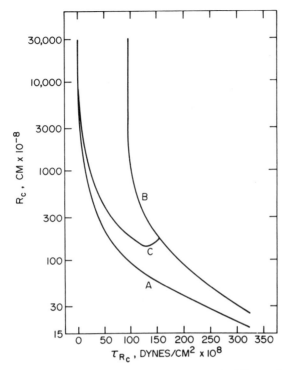

Fig. 7. Equilibrium loop radius versus applied stress curves for (A) an ordinary dislocation in a disordered lattice, (B) case shown in Fig. 5b, and (C) case shown in Fig. 5a. From Marcinkowski and Leamy[*39*].

unstable in that the innermost loop collapses and the resulting dislocation reverts to that shown by curve B. With the aid of numerical techniques, the exact energy expressions were used to determine the curves given in Fig. 7.

The energy maxima associated with the curves such as shown in Fig. 6 have an important physical significance in that they can be visualized as the activation energy necessary to nucleate a dislocation loop in an otherwise perfect crystal. The value of this activation energy is shown in Fig. 8 as a function of applied stress τ. Three cases are considered, where case C is illustrated in Fig. 9. Case A simply represents the condition of ordinary dislocation loop nucleation in a perfect disordered crystal. The nucleation of a superlattice dislocation, on the other hand, can be visualized as the two independent dislocation loop nucleation events beginning with that shown in Fig. 5b followed by the one illustrated in Fig. 9. In practice, the two nucleation events may not in fact be independent in that the second loop may nucleate in the stress field of the first[*44*], in turn raising the activation energy of the second loop. Note that in Fig. 8 curves B and D are separated

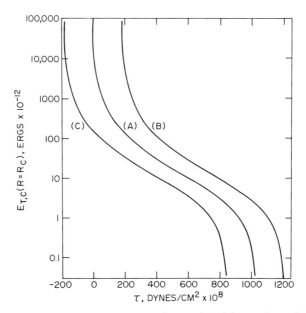

Fig. 8. The applied stress dependence of $E_{\mathrm{T,\,C}}$ $(R = R_{\mathrm{C}})$ for (A) on ordinary dislocation in a disordered lattice, (B) case shown in Fig. 5b, (C) case shown in Fig. 9. From Marcinkowski and Leamy [39].

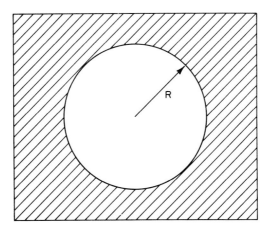

Fig. 9. Elimination of an antiphase boundary by the nucleation of a second ordinary dislocation loop.

from A by a constant stress level of $+\gamma_0/b$ and $-\gamma_0/b$, respectively. The nature of dislocation loop nucleation can lead to marked effects on the mechanical behavior of alloys as will be discussed in a later section. One of the drawbacks of the present treatment of dislocation loops is the assumption of crystalline isotropy. Similar to the case of straight dislocations treated earlier, the introduction of anisotropy is expected to have significant effects on the behavior of dislocation loops in superlattices[45]. In addition, the generation of superlattice dislocation loops from Frank–Read sources[46] may require much less energy than that of spontaneous nucleation from a perfect crystal, although the qualitative features of both processes should be similar.

C. Dislocation Interactions

The superlattice dislocations considered thus far have been isolated ones. It has been shown, however, that superlattice dislocations can perturb one another very strongly, leading to marked alterations in their respective morphologies[47]. In particular, consider the case shown in Fig. 10 where a pair of straight superlattice dislocations pass one another under the influence of the applied stress τ. For convenience, the dislocations are drawn

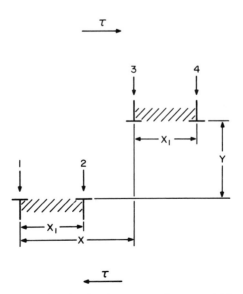

Fig. 10. Schematic illustration showing the passage of two superlattice dislocations. From Marcinkowski and Lakso[47].

with edge-type designation. It is now possible to write the total energy of the
system shown in Fig. 10 as

$$E_T = \sum E_S + \sum E_I + \sum E_{\gamma_0} + \sum \tau b \, \Delta x \qquad (26)$$

Again, this equation is very similar to that given by Eq. (1) except for the last
term. This term represents the work done by τ when each of the dislocations
moves a distance Δx measured with respect to some reference position on
the slip plane. Equation (26) is solved by ensuring that $\partial E_T/\partial x_1 = 0$ for

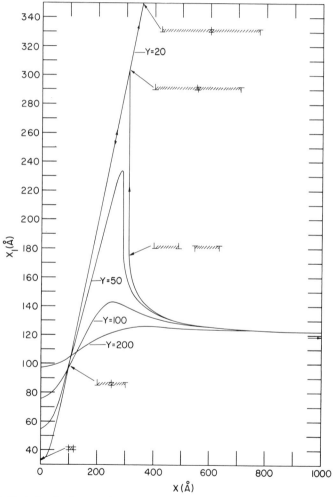

Fig. 11. Variation in the extension of a pair of superlattice screw dislocations as they pass
one another at various vertical separations Y given in angstroms. From Marcinkowski and
Lakso[47].

various values of x. The equilibrium configurations associated with Fig. 10 could also be determined using force[47] rather than energy techniques; however, the energy approach is much more general[48]. The resulting x_1 versus x curves for a number of different slip plane spacings y given in angstroms is shown in Fig. 11 for the case of the specific Fe_3Si superlattice[47]. It is immediately apparent that the extension of the super-lattice dislocations undergoes wide fluctuations as the dislocations pass one another. Finally, when the slip plane spacing becomes sufficiently small (20 Å), the mutual interaction between the superlattice dislocations causes them to become split into ordinary dislocations with the resultant genera-tion of extensive APB area. It is also possible to find the resultant stress τ associated with each value of x in Fig. 10.

D. Lattice Friction Stress

The lattice friction or Peierls stress has already been discussed in terms of dislocation mobility. Another important effect of lattice friction may be discerned with respect to Fig. 12[49,50]. The curved line represents the repulsive force exerted between the two ordinary dislocations comprising the superlattice dislocation shown in Fig. 1a. The solid line, however, corre-sponds to the force arising from the APB tension. The equilibrium extension of the superlattice dislocation corresponds to the point at which these two curves cross in Fig. 12, i.e., r_e, which is also given by Eq. (7). When a lattice

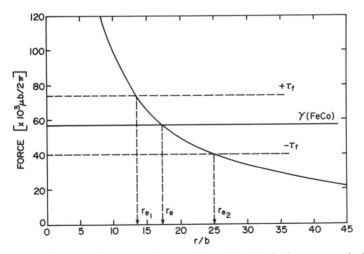

Fig. 12. Equilibrium configuration of a superlattice dislocation in the presence of a lattice frictional stress. From Sadananda and Marcinkowski[50].

friction stress τ_f is present, the above results are significantly modified, depending upon how the superlattice dislocation is formed. If the two dislocations are relaxed from a closely separated condition, equilibrium is attained when the repulsive force between the dislocations is just balanced by the sum $(\gamma_0 + \tau_f b)$ leading to the equilibrium separation r_{e_1} in Fig. 12. If, on the other hand, the dislocation pair is allowed to relax from a large distance it is attracted by the force $(\gamma_0 - \tau_f b)$ and attain equilibrium at r_{e_2}. A stable equilibrium is attainable anywhere between r_{e_2} and r_e. If, on the other hand, the frictional stress is greater than γ_0, the upper limiting value r_{e_2} goes to infinity. The specific calculations illustrated in Fig. 12 were carried out for the specific case of the fully ordered FeCo superlattice, where τ_f was taken as the yield stress. It is immediately apparent from the above analysis that great care should be exercised in calculating γ_0 by means of Eq. (7) from values of r observed within the electron microscope when the lattice friction stress is high or when γ_0 is small.

Lattice friction can act in two distinct ways[51,52]. This can be seen with reference to Fig. 10 by writing the expression for the total force equilibrium on one of the individual dislocations, say dislocation 3, as the array moves. In particular

$$F_3 = F_I \pm F_{I,f} + F_{\gamma_0} + F_a - F_f = 0 \tag{27}$$

The first term in the above equation is simply the total elastic interaction force between dislocation 3 and all of the others. The second term represents the frictional force which balances the internal stress given by the first term and the APB energy given by the third term, as discussed in connection with Fig. 11. This second term exists whether the dislocation is moving or remains static. The fourth term in Eq. (27) arises from the applied stress and is given by τb; the last term is due to the friction stress acting on a moving dislocation, i.e., $\tau_f b$. The last term does not exist for a static dislocation.

The introduction of a lattice friction stress also significantly modifies the x_1 versus x curves shown in Fig. 11 by introducing a marked hysteresis into these relationships, depending upon whether x is increasing or decreasing[50]. Up to the present time, calculations have not yet been carried out with respect to the passage of coaxial superlattice dislocation loops with or without the presence of friction. Calculations of this type, however, have been performed with ordinary dislocation loops and show that the passing stress between loops of opposite sign increases markedly as the loop radius is decreased[53]. This is due to much the same reason as that encountered in connection with Eqs. (23) and (25), namely, the approximately inverse relationship between τ and R for circular dislocation loops. Similar calculations with superlattice dislocations should also prove to be of interest, especially with respect to uncoupling effects. It is also important to note that the concept of friction has heretofore been discussed with respect to lattice friction, although the friction could be of a much more general nature. In particular, it could arise from dislocation–impurity, dislocation–solute, etc. interactions.

III. Effect of Atomic Order on Flow Stress

Having discussed the various dislocation morphologies to be anticipated in ordered alloys, it is of importance to examine next a number of experimental results relating to the effect of atomic order on mechanical properties and to reconcile the two. Figure 13 shows the stress–strain curves associated with a fully ordered FeCo alloy over a range of low temperatures which do not affect the state of atomic order[54]. The curves may be divided into three well-defined regions or stages. Stage I may be associated with the yield point which is seen to be rather sharply defined. This stage is followed

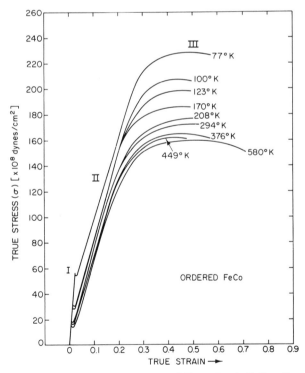

Fig. 13. Stress–strain curves obtained from fully ordered FeCo alloys at various temperatures. From Fong *et al.*[54].

by a region of high linear work hardening, i.e., large $d\sigma/d\varepsilon$ labeled as stage II in Fig. 13. Following stage II is a region of nearly zero work hardening. When the same FeCo alloy is disordered by rapid quenching, the stress–strain curves shown in Fig. 14 are obtained[54]. These curves are seen to differ markedly from those of the corresponding ordered alloys. In particular, the sharp yield point as well as stage II are no longer present. For convenience, however, the yield point for these alloys will be described in terms of the flow stress at a plastic strain of 0.002. The disordered FeCo alloy can thus be visualized as going into stage III immediately upon yielding.

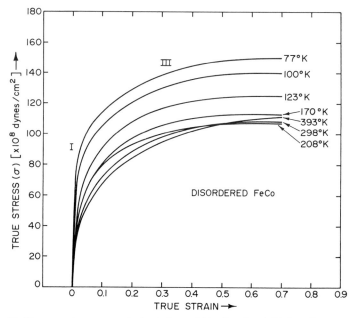

Fig. 14. Stress–strain curves obtained from fully disordered FeCo alloys at various temperatures. From Fong *et al.*[54].

It is of interest to examine next the yield stresses (stage I) associated with the FeCo alloy as a function of both quench temperature and elevated temperature after equilibrium degrees of atomic order were attained. The results are shown in Fig. 15, where the quenched samples were tested at room temperature ($\simeq 298°$K)[55]. Similar curves have been obtained by other investigators with the FeCo–2V alloy[56,57]. The very high values of the yield stress above T_C for the quenched samples have been associated with the formation of a very high degree of short-range order during the quench, and perhaps even some long-range order. The reason for this can be seen in Table II where the ratio of the critical ordering temperature to the melting

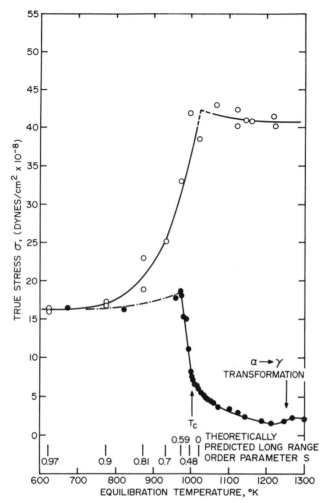

Fig. 15. Comparison of the as-quenched room temperature and elevated temperature yield stress of equilibrated FeCo alloys. From Marcinkowski and Chessin[55]. ○ Flow stress in compression at $\varepsilon_p = 0.002$ measured at room temperature after quenching from the temperature indicated; ● flow stress in tension at $\varepsilon_p = 0.002$ measured at temperature indicated.

point of this alloy T_C/T_M possesses the relatively high value of 0.59, so that the freezing in of the equilibrium degrees of order by quenching in the temperature range near T_C is very difficult. If, on the other hand, similar quenching experiments are carried out with the Ni_3Mn alloy in which T_C/T_M has the relatively low value of 0.504, the results shown in Fig. 16 are obtained[58]. The yield stress may be taken as the bottommost curve in this figure. It is immediately apparent that the fully ordered alloys possess a somewhat higher yield stress than those quenched from above T_C, while those with intermediate states of ordering exhibit a pronounced strengthening. Contrary to the quench data observed in Fig. 15, the results shown in Fig.16 are believed to represent very nearly equilibrium states of order for the temperatures indicated. These findings are also in general agreement

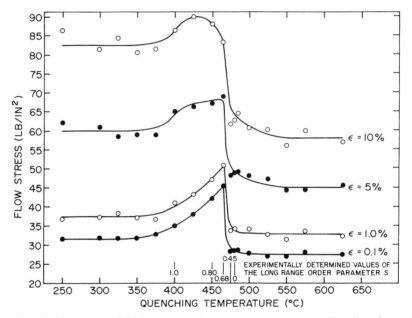

Fig. 16. Flow stress of Ni_3Mn measured at room temperature as a function of quench temperature for the plastic strains indicated. From Marcinkowski and Miller[58].

with subsequent studies[59,60]. Results showing the same general features as those observed for FeCo and Ni_3Mn have also been observed for Fe_3Al[56,61] as well as for CuZn, i.e., β-brass[62–64]. The flow stress and work-hardening rates are shown as a function of quenching temperature in Fig.17 for this latter alloy[62].

A number of theories have been advanced to account for the above results. In the first place, it was recognized that at temperatures low compared with T_C, dislocations will usually be present as superlattice dislocations, whereas near and above T_C ordinary dislocations will represent the more stable configuration. This argument was used to account for the maximum in the yield stress–temperature curves[56]. On the other hand, it has been postulated that superlattice dislocations should also exist above T_C as a result of

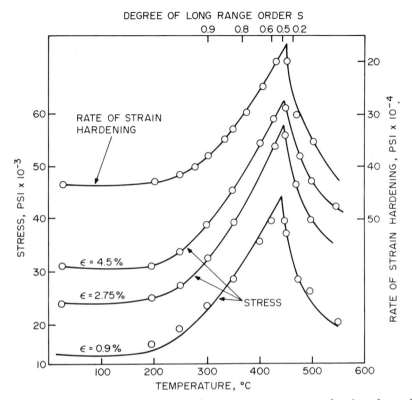

Fig. 17. Flow stress of β-brass measured at room temperature as a function of quench temperature for the plastic strains indicated. From Green and Brown[62].

short-range order[65,66]. Indeed, TEM techniques have shown dislocation pairs to exist in short-range ordered Cu_3Au[67], FeCo–2V[57] and Mg_3Cd[68] as well as in some austenitic stainless steels and α-brasses[69]. Figure 18 shows the results of a number of calculations[56,57,70] relating the superlattice dislocation extension in FeCo to the equilibrium temperature[57]. The results of Moine *et al.*[57] in Fig. 18 were obtained by using Eq. (7) where γ_0 was given by[71]

$$\gamma_{(hkl)} = \frac{\sigma^2 + S^2 - 3S^2\sigma + \sigma\, NZv}{1 - S^2}\,\frac{}{2} \tag{28}$$

where S and σ are the long- and short-range order parameters[6,19], N the number of atoms per unit area on the hkl plane, and Z the number of bonds per unit area across the hkl plane. For the (110) plane, $N = \sqrt{2}/a_0^2$ while $Z = 2$. According to Moine *et al.*[57] the yield stress is determined by the stress to operate a Frank–Read source. In particular, at temperatures well

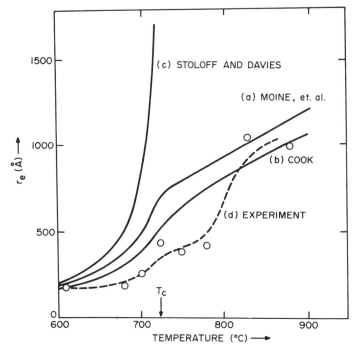

Fig. 18. A comparison of the equilibrium extension of superlattice dislocations in FeCo–2V as a function of temperature from a number of different theories. From Moine *et al.*[57].

below T_C, this stress is given by Eq. (23), whereas near and above T_C, the yield stress is given by Eq. (25). The results based on these equations are shown in Fig. 19 where they are compared with the experimental yield stresses obtained for the FeCo alloy as a function of quench temperature. For the theoretical results, R was taken as 3500 Å, and is thus an arbitrary parameter. The justification for this value is not obvious.

Equation (25) is based upon the argument that since the superlattice dislocations are weakly coupled, they behave nearly independent of one another. In this sense, therefore, the theories of Stoloff and Davies[56] and Moine *et al.*[57] are similar, and both predict a maximum in the yield stress at some intermediate degree of long-range order. Physically, it is a simple matter to account for the maximum in that at very high temperatures there should be no order strengthening, while at very low temperatures the order should be perfect and superlattice dislocations would again move through the lattice with no resistance due to atomic order. It is at the intermediate values of order that it becomes impossible for any type of dislocation, either superlattice or ordinary, to move through the crystal without creating disorder.

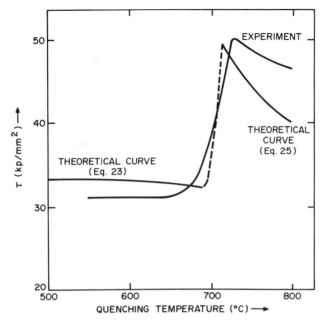

Fig. 19. Comparison of the theoretical and experimental results of the yield stress as a function of quench temperature for the FeCo–2V superlattice. From Moiné *et al.*[57].

Rudman[72] was the first to realize that even for superlattice dislocations, the contribution from σ would give rise to order strengthening for $0 < S < 1$.

An alternative expression for Eq. (27) was first given by Flinn[73] based on an earlier model for short-range-order strengthening first formulated by Fisher[74]. In this formulation it was assumed that an ordinary dislocation completely randomized, i.e., destroyed, the short-range order across the slip plane. It was found that for a body centered cubic lattice

$$\gamma_{(110)} = (8m_A m_B v\alpha_1)/\sqrt{2}\,a_0^2 \tag{29a}$$

where m_A and m_B are the mole fractions of the two components in the alloy and α_1 is the nearest neighbor short-range order parameter defined as

$$\alpha_1 = 1 - (P_{AB}/m_A) \tag{29b}$$

where P_{AB} is the probability of finding a particular A atom as a nearest neighbor to a B atom. According to Fisher[74], the stress τ required to move an ordinary dislocation with Burgers vector $b = \frac{1}{2}a_0\sqrt{3}$ against the surface tension $\gamma_{(110)}$ is simply $\gamma_{(110)}b$ giving

$$\tau = 16m_A m_B v\alpha_1/\sqrt{6}\,a_0^3 \tag{30}$$

The above result can also be obtained from Eq. (25) as $R \to \infty$. In addition, above T_C, $S = 0$, and Eq. (28) reduces to that of Eq. (29a) for the particular case where $m_A = \frac{1}{2}$ since here $\sigma \simeq \alpha_1$. On the other hand, for large values of S, $\sigma \simeq S^2$ and Eq. (28) reduces to

$$\gamma_{(110)} = 4vS^2/a_0^2\sqrt{2} \tag{31}$$

in agreement with earlier findings[*11*]. Thus, except for the dislocation loop nature of Moine *et al.'s*[*57*] calculations, they are conceptually the same as those of Flinn's[*73*] straight dislocation model.

In spite of the similarity in strengthening behavior between β-brass and the other ordered alloys discussed thus far, Brown[*75*] has argued that in the former case the effects are due to quenched-in vacancies. It is further argued that it is not possible to quench in any significant amounts of disorder due to the high value of $T_C/T_M = 0.643$ listed in Table II. A theory has also been presented for the elevated temperature yield stress results for β-brass based upon the difference in energy between a diffusion smoothed APB and a sharp slip-produced APB[*63,64*]. Such a theory assumes that the dislocations responsible for yielding are already present within the crystal before yield. This assumption, however, appears to be incompatible with the Frank–Read or loop generation mechanism. Sumino[*76*] has also formulated a theory of order strengthening based upon stress-induced changes in the degree of atomic order in the vicinity of a dislocation due to the stress field of the dislocation itself. Such a theory is clearly insufficient to account for the results obtained with the quenched samples where the test temperatures are sufficiently low so as to preclude atomic diffusion.

Referring again to Figs. 15 and 16 shows that the yield stresses for the equilibrated fully ordered alloys is somewhat higher than that for the equilibrated disordered alloys. Similar findings have been made with respect to the Fe_3Al alloy[*56,61*]. Since superlattice dislocations should move through a fully ordered alloy with very little resistance, it follows that all of the dislocations responsible for the deformation in these alloys may not be of the superlattice type. Dislocation produced APBs have been observed in fully ordered Cu_3Au[*77*] and Fe_3Al[*78*] which have been subjected to relatively low stresses, and would, therefore, seem to substantiate the above postulate. More will be said concerning this point in later sections.

Consideration of Table I shows that the B2 type NiAl[*33*], AgMg[*79,80*], and AuZn[*28*] alloys have $\langle 100 \rangle$ as their slip direction except for orientations close [001] in which case the slip direction becomes $\langle 111 \rangle$. The reason for the choice of $\langle 111 \rangle$ over $\langle 100 \rangle$ for orientations close to [001] is due to the fact that the resolved shear stress associated with $\langle 100 \rangle$ for these orientations approaches zero. Further consideration of Table II shows that these particular alloys are ordered up to their melting point, i.e., $T_C/T_M = 1$.

Because of the resulting high APB energies, the superlattice dislocations with Burgers vectors along $\langle 111 \rangle$ are virtually undissociated according to Eq. (12) so that dislocations with a [100] Burgers vectors become more stable. Furthermore, because of the coincidence of T_C with the melting point in the above-mentioned B2 type alloys, none of the intricacies in flow stress associated with the order–disorder transition such as shown in Figs. 15, 16, and 17 are present in these particular alloys.

Thus far only ordered alloys possessing cubic crystal structures have been considered, and these constitute the majority of known superlattice alloys. Two noncubic type superlattices are shown in Tables I and II. They are the $D0_{19}$ type structure, which is hexagonal, and $L1_0$ type, which is tetragonal. The most widely studied of the $D0_{19}$ type structures is Mg_3Cd. Its yield stress exhibits many of the same features as that shown by the FeCo alloy in Fig. 15[*81*]. However, whereas basal slip is the predominant mode of plastic deformation for the disordered alloy, prismatic slip becomes favored for the ordered alloy[*68,81,82*]. The most probable explanation for this behavior can be seen by reference to Fig. 20[*68*]. Figure 20a shows the atom positions on two successive layers of the basal plane and the dislocations on these planes are essentially the same as in the close packed $L1_2$ type structures[*11*]. On the other hand, Fig. 20b shows the projections of the corrugated $(01\bar{1}0)$ prism and $(01\bar{1}1)$ pyramidal planes. Two types of planes may be associated with each of these, and they are labeled as type 1 and type 2 in Fig. 20b. In the latter case, the APBs comprising the superlattice dislocations are of NN type[*68*], while those in the former case are of NNN type[*11,68*] and are thus of relatively low energy. The superlattice dislocations lying on type 1 planes will in turn be of correspondingly low energy in accordance with Eq. (5) or (11). Thus, it seems reasonable to assume that for the disordered case, it is the low value of the stacking fault energy that favors the dislocations to lie in the basal planes[*83*], whereas for the ordered alloys, it is the low value of the APB energy that favors the dislocations to lie in the prism planes.

The fifth and final class of superlattice alloys to be considered in the present review is the tetragonal $L1_0$ type. Two kinds of superlattice dislocations are predicted to exist in this particular superlattice type and are shown in Table I[*11,15*]. Except for the addition of the NN APB, the dislocation configuration consisting of a pair of partials is identical to that existing in face centered cubic lattices. To date, the dislocation configurations predicted in Table I for the $L1_0$ superlattices has not been verified experimentally. This is in part due to the complexity of the ordering transformation. In particular, above T_C the corresponding disordered alloy is cubic. The tetragonality induced by ordering gives rise to severe internal strains which are relieved by twinning. Such twinning has been verified in the AuCu[*84–86*], CoPt[*87–89*], and FePt[*90*] alloys. A theory has been presented to account for the manner

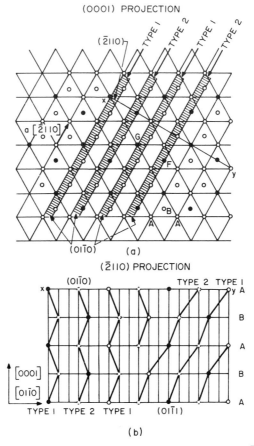

Fig. 20. (a) (0001) projection of the Mg$_3$Cd lattice. The corrugated (01$\bar{1}$0) planes are also shown in projection. (b) ($\bar{2}$110) projection of the Mg$_3$Cd lattice showing the corrugated (01$\bar{1}$0) and (01$\bar{1}$1) planes. From Blackburn[68]. ● Cadmium atoms; ○ magnesium atoms; ◌ magnesium atoms not lying in plane of projection; atoms labeled A & B for hexagonal stacking sequence ABABAB.

in which twinning can relieve the strains occasioned by the tetragonal distortions[91] and a dislocation model has also been suggested for the twinning mechanism[92]. Although it has been argued that subsequent deformation of ordered CuAu takes place by dislocation motion[93,94], this has been disputed by Pashley et al.[95] who argue that such deformation occurs by means of deformation twinning. Flow stress temperature measurements similar to those shown in Figs. 15, 16, and 17 seem not to have been obtained for the L1$_0$ type superlattices. Before proceeding further, it should be mentioned that the APB energies given in the last two columns of Table II were all obtained from TEM observations using equations of the type given by (7) or (12) depending upon whether or not the single crystal elastic constants are known. The numbers in parentheses next to each value of γ_{NN} or γ_{NNN} refer to the specific reference from which the value was obtained.

A. Time–Temperature Effects

When an ordered alloy is rapidly quenched from above T_C so as to retain a high degree of disorder and subsequently annealed for various times at some constant temperature below T_C, the yield stress passes through a maximum value. This behavior can be seen in Fig. 21 for a single crystal of Cu_3Au[99],

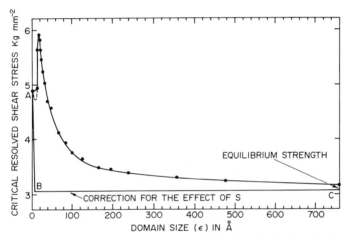

Fig. 21. Room temperature yield stress versus antiphase domain size for a single crystal of Cu_3Au ordered at 350°C. From Ardley[99].

where the yield stress was measured at room temperature after pulse anneals at 350°C for various times followed by rapid quenching. The abscissa is given in terms of antiphase domain size (which in turn is a function of time) since, as will be seen shortly, the antiphase domain size is thought, at least in one theory, to be responsible for the maximum in the strengthening. Similar results have been observed for Fe_3Al[61,100,101], Ni_3Mn[58–60,102], Ni_3Fe[59,60,102,103], Mg_3Cd[81], CuAu[102,104,105], and CuAu[106–108]. In the case of CuAu there is some disagreement as to whether a maximum exists in the flow stress–time curves[109–111], however, this seems to be dependent upon temperature, occurring only at the higher temperatures. Although the flow stress of β-brass decreases with time after quenching from above T_C[62], no maximum is observed. As discussed earlier, Brown[75] has attributed this to vacancy effects. On the other hand, the possibility cannot be ruled out that a domain size effect exists, and that the failure to observe a maximum may be due to the rapid growth of domains both during and after the quench from above T_C, but prior to testing.

The strengthening contribution from APBs was first calculated by Cottrell[22]. In particular, reference to Fig. 22 shows the effect of passing a single superlattice dislocation through the vertical APB. Two horizontal APB ledges, each of strength $2b = 2a$, are produced. When N such superlattice dislocations are passed through the APB such that $N(2b) = l/2$ the slip

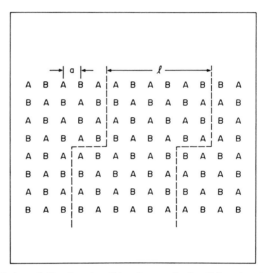

Fig. 22. Disordering of slip plane by glide of a superlattice dislocation through antiphase boundaries.

plane can be considered as completely disordered, i.e., just as many nearest neighbor A—A and B—B bonds exist as A—B bonds. The energy associated with this disordered slip plane can then be written as $\gamma_0/2$. Assuming that before any glide had occurred the antiphase domains were in the form of cubes, then the volume fraction of material comprising the APBs is given as $6l^2a/l^3$. Since this APB is shared by two neighboring domains, its energy can be written as $(6l^2a/l^3)\gamma_0/2$. In the case of an arbitrarily shaped domain we may write $(\alpha l^2a/l^3)\gamma_0/2$, where α is a geometrical form factor. The complete disordering of the slip plane by superlattice dislocations can thus be written as

$$\frac{\gamma_0}{2}\left(1 - \frac{\alpha a}{l}\right) = \tau\left(\frac{l}{2}\right) = \tau N(2b) \tag{32}$$

where the leftmost side of the above equation is simply the difference in APB energy on the slip plane before and after glide, while the two τ terms on the right represent the external work from the applied shear stress τ necessary to disorder the slip plane. Equation (32) can be rearranged to give

$$\tau = \frac{\gamma_0}{l}\left(1 - \frac{\alpha a}{l}\right) \tag{33}$$

which has a maximum value of

$$\tau_{\max} = \gamma_0/4\alpha a \tag{34}$$

at a domain size of $l = 2\alpha a$. A number of modifications of Eq. (33) have been made[99,112], however they are all basically the same.

An alternative hypothesis exists for explaining the maximum in the curve of Fig. 20[102] and is similar to that already used in Figs. 15, 16, and 17. Specifically for times near $t = 0$ when S is close to zero, the dislocations behave as ordinary dislocations, while for larger times when $S \to 1$, the dislocations move as superlattice dislocations. In the former case the yield stress increases with time, while in the latter case it decreases with time. A maximum in yield stress is obtained corresponding to the transition from ordinary to superlattice dislocations. It should be emphasized at this point that during the early stages of isothermal ordering both domain size as well as the degree of order within the domains is increasing. In this respect, it would be more desirable to study small domain size effects by investigating the mechanical properties of long-period superlattices. Such structures have been observed in both the Cu_3Au[11,67] as well as CuAu[11,113] alloys, and have the advantage in that they are well-defined equilibrium configurations.

B. Temperature Dependence of Flow Stress for Fixed States of Order

When the critical ordering temperature T_C is sufficiently low, i.e., $T_C/T_M \leq 0.60$, it is possible by suitable heat treatments to prepare nearly fully ordered and fully disordered states within the same alloy. The corresponding flow stresses can then be subsequently examined as a function of temperature over a low temperature range that does not alter the fixed states of order. Results of this type are shown in Fig. 23 for the FeCo alloy[114]. It can be seen that over most of the temperature range, the yield stress for the ordered alloy is less temperature dependent than that for the disordered alloy. Similar results have also been obtained with FeCo–2V[115]. The lowest-temperature liquid-helium (4.2°K) results for the ordered and disordered alloys correspond to deformation via crack and twin formation and will be discussed more fully in a later section. The lower temperature dependence of the yield stress for ordered alloys compared to their corresponding disordered states appears to be a general phenomenon and has been observed

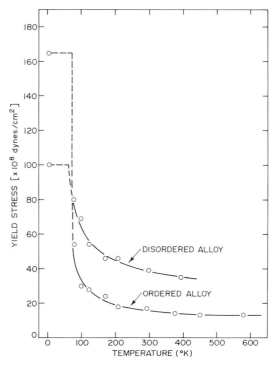

Fig. 23. Temperature dependence of the yield stress for ordered and disordered FeCo. From Fong *et al.*[114].

in Cu$_3$Au[116–119], Mg$_3$Cd[120,121], Fe$_3$Si[34], and Fe$_3$Al[44]. Figure 24 shows the temperature dependence of the critical resolved shear stress in Mg$_3$Cd single crystals for slip on both the prismatic and basal planes[121]. The dramatic decrease in temperature dependence of this stress with atomic order is immediately apparent.

The low temperature dependence of the yield stress has generally been attributed to the lattice friction or Peierls stress[122]. It therefore becomes important to know the manner in which atomic ordering reduces this friction stress. It has generally been recognized that a dislocation with Burgers vector $\frac{1}{2}a_0[111]$ in a body-centered cubic lattice can lower its energy by a dissociation on the $(11\bar{2})$ plane of the type given by[37]

$$\tfrac{1}{2}a_0[111] = \tfrac{1}{6}a_0[111] + \tfrac{1}{3}a_0[111] \tag{35}$$

Furthermore, complex dissociations given by[123,124]

$$\tfrac{1}{2}a_0[111] = \tfrac{1}{6}a_0[111] + \tfrac{1}{6}a_0[111] + \tfrac{1}{6}a_0[111] \tag{36}$$

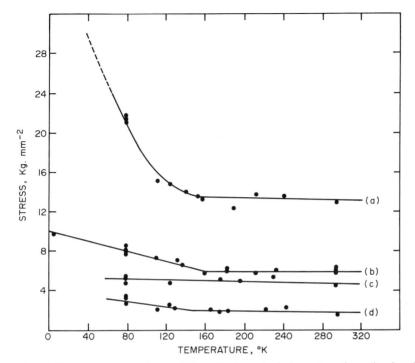

Fig. 24. Temperature dependence of the critical resolved shear stress for ordered and disordered Mg_3Cd single crystals. From Noble and Kirby[121]. (a) Prismatic slip, $S = 0.0$. (b) prismatic slip, $S = 1.0$; (c) basal slip, $S = 1.0$; (d) basal slip, $S = 0.0$.

and

$$\tfrac{1}{2}a_0[111] = \tfrac{1}{8}a_0[101] + \tfrac{1}{8}a_0[110] + \tfrac{1}{8}a_0[011] + \tfrac{1}{4}a_0[111] \qquad (37)$$

have also been proposed, where the reactions shown in Eq. (36) occur on {112} type planes, while those in Eq. (37) take place on planes of the type {110}. The dissociations given by Eqs. (36) and (37) are illustrated schematically in Figs. 25a and 25b, respectively. Other reactions can be postulated but they are simply variations of the ones listed above.

Since the stacking fault energies associated with the dislocation dissociations in Fig. 25 are assumed to be large, the separation between the partials is expected to be small. These separations, however, are sufficiently large so as to make the entire extended dislocation configuration sessile with respect to glide. In order for glide to occur, the partials must be recombined into a single $\tfrac{1}{2}a_0[111]$ type dislocation and/or become extended only on the slip plane. This can be done by a suitable combination of applied stress and thermal activation[124], and presumably gives rise to the large temperature dependence associated with glide in body-centered cubic structures. It is

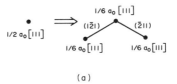

(a)

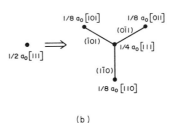

(b)

Fig. 25. Two possible types of dislocation dissociation in a body-centered cubic lattice.

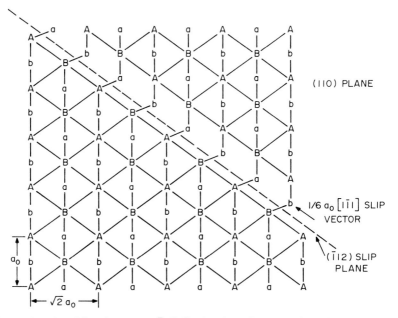

Fig. 26. Creation of disorder across a $(\bar{1}12)$ slip plane in a B2 type superlattice by the passage of a partial dislocation with Burgers vector $\frac{1}{6}a_0[1\bar{1}1]$. A, B are atoms in plane of drawing; a, b atoms in plane immediately below and above that of drawing; ———— first nearest-neighbor bonds.

therefore important to know what effect atomic ordering has on the extended dislocation configurations shown in Fig. 25. This can be most readily seen by reference to Fig. 26 which shows the effect of the passage of a $\frac{1}{6}a_0[1\bar{1}1]$ partial dislocation across the $(\bar{1}12)$ slip plane (shown dashed) in a B2 type superlattice. It is immediately apparent that this glide process gives rise to incorrect nearest neighbor atom pairs across the glide plane. It also follows from the geometry of Fig. 26 that the energy expended in disordering the $(\bar{1}12)$ glide plane is $4v/(\sqrt{6}\,a_0^2)$. It can be shown that a similar disordering results from the motion of a $\frac{1}{8}a_0[110]$ type partial dislocation on the $(1\bar{1}0)$ plane[125]. Thus, atomic ordering is expected to lead to a marked decrease in the dissociation of the extended dislocations shown in Fig. 25 with a consequent reduction in the Peierls stress and a corresponding reduction in the temperature dependence of this stress, in agreement with the observed experimental findings. It will also be shown in a future section that the slip process illustrated in Fig. 26 is intimately related to deformation twinning in body-centered cubic and B2 type lattices.

The flow stress of ordered alloys can vary strongly with temperature as a result of processes other than the Peierls stress. This can be seen by reference to Fig. 27 which shows the temperature dependence of the yield stress (solid circles) for a fully ordered Fe_3Si superlattice[34]. Between B and B' the yield stress is essentially temperature independent, as anticipated from the previous discussion. Between B and A', however, there is a marked increase in the yield stress with decreasing temperature. This rapid increase has been associated with the difficulty of nucleating superlattice dislocations at these low temperatures. In particular, the dislocations are nucleated as pairs of

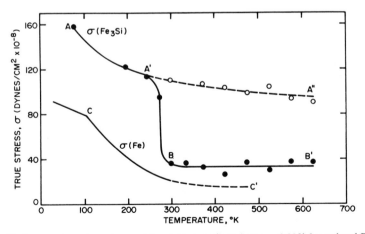

Fig. 27. Temperature dependence of the yield stress (based on $\varepsilon = 0.002$) for ordered Fe_3Si and pure iron.

$\frac{1}{4}a_0\langle 111 \rangle$ type ordinary dislocations, which in turn generate NNN APBs, as can be seen by reference to Table I. These dislocation generated NNN APBs can be observed in the dark field transmission electron micrograph of Fig.28 which was obtained from a fully ordered Fe_3Si single crystal strained at $243°K$.

The reason for difficulties associated with the dislocation nucleation process in ordered alloys can be most easily visualized by reference to Figs. 6 and 8 where it is seen that a certain activation energy is necessary to nucleate each of the individual ordinary loops comprising the perfect superlattice dislocation. The strain rate $\dot{\varepsilon}$ associated with the deformation of a superlattice controlled only by the nucleation process may be written as[39]

$$\dot{\varepsilon} = nfab \exp \sum_i E_i/kT \qquad (38)$$

where n is the number of nucleation sites, f is a frequency factor associated with the vibrational spectra that give rise to the nucleation event, a is the area swept out by each loop, and E_i the energy associated with nucleation of each ordinary dislocation loop comprising the superlattice dislocation and which must be supplied by thermal fluctuations within the crystal. It is apparent that at sufficiently low temperature the thermal energy within the lattice may not be sufficient to nucleate all of the ordinary dislocations to form a perfect superlattice dislocation, so that imperfect superlattice dislocations with their consequent production of APBs will be generated. This APB tension will in turn raise the yield stress of the crystal by the amount γ_0/b above that of the already existing Peierls stress, as is borne out quantitatively in the case of the Fe_3Si alloy[34].

The open circles associated with the dashed curve A'–A" in Fig. 27 represent the flow stresses corresponding to the beginning of stage III. This curve is seen to be simply an extension of the yield stress curve A–A' associated with the low-temperature data. The curve AA'A" is thought to represent the same temperature dependent deformation process, and since it roughly parallels that for pure iron, i.e. curve C–C', it is presumed to be identical to the same temperature-dependent deformation processes occurring in pure iron. It seems reasonable to attribute this temperature dependence to the Peierls or friction stress. It is also of interest to examine the temperature dependence of the flow stress associated with stage III deformation in other ordered and corresponding disordered alloys. In particular, Fig. 29 shows the temperature dependence of the flow stress for fully ordered FeCo at the various indicated plastic strains[54,55]. Similarly, Fig. 30 shows the temperature dependence of the flow stress for the corresponding disordered FeCo alloys at selected plastic strains[54,55]. The temperature dependence of the flow stress for the disordered alloys in Fig. 30 is seen to be essentially the same for

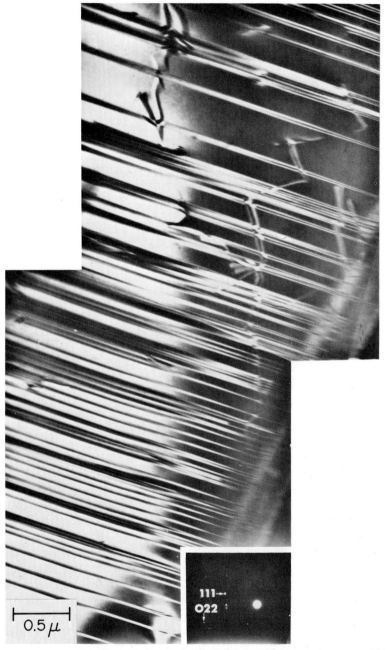

Fig. 28. Transmission electron micrograph showing slip produced second nearest-neighbor type antiphase boundaries in fully ordered Fe$_3$Si deformed at 243°K. From Lakso and Marcinkowski[34].

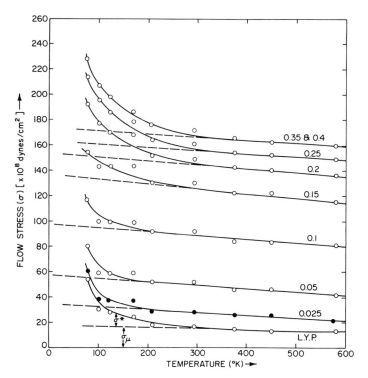

Fig. 29. Temperature dependence of the flow stress for ordered FeCo at various plastic strains. From Fong *et al.*[*54*].

all strains, and is thus thought to correspond to the same process, namely the overcoming of the Peierls barrier. The upward rise of the 0.65 strain curve in Fig. 30 with increasing temperature is associated with the point defect induced reordering which is enhanced by large plastic strains and high temperatures. Reference to the flow stress–temperature curves in Fig. 29 for high strains, i.e. stage III, shows them to be nearly the same form as those for the corresponding disordered alloys, again indicating similar processes. It can thus be concluded that the relatively weak temperature dependence of the flow stress associated with stages I and II of the fully ordered FeCo alloy is associated with the movement of less extended ordinary dislocations (in the sense shown in Fig. 25) comprising the superlattice dislocation over the Peierls barriers. On the other hand, the stronger temperature dependence of the flow stress characteristic of stage III of the ordered alloys and the entire range of strains (stage III) associated with the corresponding disordered alloys is related to the movement of more fully extended ordinary disloca-

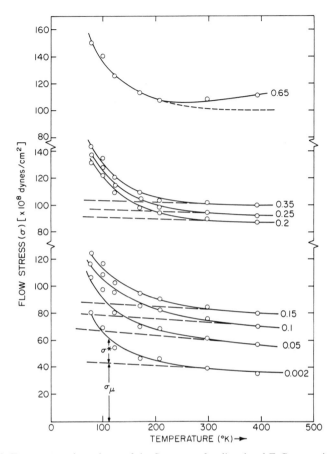

Fig. 30. Temperature dependence of the flow stress for disordered FeCo at various plastic strains. From Fong *et al.*[54].

tions (in the sense shown in Fig. 25) over the Peierls barriers. It is implied in this latter statement that the fully ordered alloy becomes highly disordered in stage III and that the dislocations begin to move, for the most part, as ordinary dislocations rather than as superlattice dislocations[55]. Similar conclusions have been reached with respect to stage III deformation in ordered Ni$_3$Al[126]. A possible reason for this behavior is the uncoupling of superlattice dislocations by the mechanisms discussed in connection with Figs. 10 and 11 which can occur at high dislocation densities, i.e. high plastic strains. That appreciable disordering of an ordered alloy by plastic deformation does indeed occur can be seen by reference to Figs. 31 and 32 which

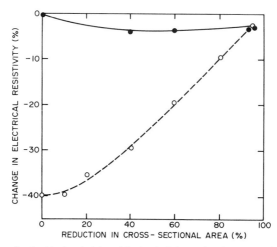

Fig. 31. Change in electrical resistivity with plastic deformation in ordered (broken line) and disordered (solid line) Ni_3Mn. From Dahl[127].

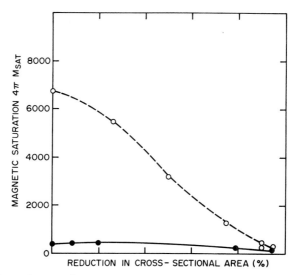

Fig. 32. Change in saturation magnetization with plastic deformation in ordered (broken line) and disordered (solid line) Ni_3Mn. From Dahl[127].

show the increase in electrical resistivity and the decrease in magnetization, respectively, brought about by plastic deformation in the ordered Ni_3Mn alloy[127]. In both cases a 100% reduction in area changes both of these physical properties to values identical to that of the corresponding disordered alloy. The state of order in both cases can therefore be assumed to be essentially the same, i.e. fully disordered. Studies with single crystals of

Ni$_3$Mn have also produced parallel results[128]. Similar findings have also been made with respect to electrical resistivity measurements obtained from cold-worked ordered alloys of Cu$_3$Au[129–131] and Cu$_3$Pd[129], as well as by X-ray measurements carried out with deformed alloys of ordered Cu$_3$Pt[132] and Cu$_3$Au[133]. Mössbauer studies have been used to detect the presence of disorder in deformed FeAl alloys[134]. More refined measurements have also been carried out with ordered Ni$_3$Mn, in which it has been shown that the maximum changes in resistivity, and presumably atomic order, occur in stage III of the deformation[135,136].

It has already been seen in Tables I and II that a large number of B2 type superlattices in which $T_C/T_M = 1.00$ have $a_0\langle 100\rangle$ as a slip or Burgers vector, along with $a_0\langle 111\rangle$, depending upon orientation with respect to the applied stress. In view of the possibility that the temperature dependence associated with the critical resolved shear stresses for both these slip vectors may be different, it follows that the temperature dependence of the yield stress obtained from polycrystals may be indeterminate since in these cases both slip systems are expected to be operative. Polycrystalline temperature dependent deformation studies of this type have been carried out for NiAl[137–140], AuZn[141], and AgMg[142].

Perhaps the most extensive study of a single crystalline B2 type superlattice with $T_C/T_M = 1.00$ has been with AgMg[79,80]. Figure 33 shows the

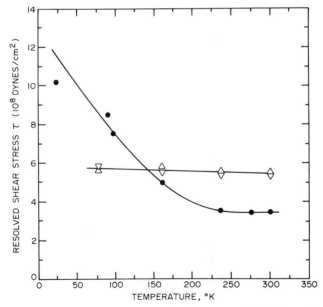

Fig. 33. Temperature dependence of the resolved shear stress in single crystals of ordered AgMg. From Mitchell et al.[79]. ∇, $\triangle$ $(hk0)$ [001] slip system; ● $(1\bar{1}2)$ [$\bar{1}11$] slip systems; strain rate $\gamma = 1.67 \times 10^{-5}$/sec.

temperature dependence of the resolved shear stress in single crystals of this alloy tested in tension for both the $\langle 111 \rangle$ and $\langle 100 \rangle$ slip directions. It can be seen that there is practically no temperature dependence of the critical resolved shear stress for slip along [001], whereas that associated with glide along [$\bar{1}$11] is appreciable. The above findings are in agreement with earlier results[143] which have been reanalyzed in reference 79 as well as with those findings obtained by bending[144]. It is also important to note that all single crystals of AgMg deformed in compression slipped along $\langle 111 \rangle$, whereas in tension, only those orientations toward the [001] corner of the unit stereographic triangle deformed in this manner. Tensile tests performed on single crystals of AuZn[28] also seem to be in agreement with the AgMg findings with respect to slip mode, although no temperature dependence of the yield stress has yet been determined for this particular alloy. The single crystalline results obtained with NiAl seem to be dramatically at odds with the AgMg and AuZn findings in that for compressive deformation, NiAl deforms along $\langle 111 \rangle$ for orientations near [001], and along $\langle 100 \rangle$ for all other orientations of the compression axis[27,33,137,138]. Furthermore, the critical resolved shear stress for $\langle 100 \rangle$ glide shows a strong temperature dependence[137,138].

In order to understand the above orientation dependence of the slip systems in B2 lattices with large values of T_C/T_M, consider first the single crystal data obtained for niobium and shown in Fig. 34[123]. The slip

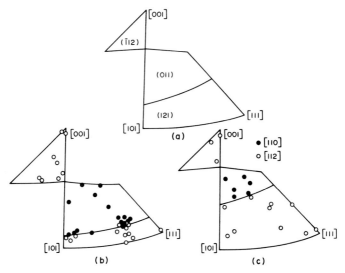

Fig. 34. Orientation dependence of the active slip planes in niobium single crystals deformed at 295°K: (a) theoretical, (b) tension, (c) compression. From Foxall *et al.*[123].

vector associated with all of these slip systems is [1$\bar{1}$1]. For reference purposes, Fig. 34a shows the operative slip planes upon which the resolved shear stress is greatest, assuming that the critical resolved shear stress associated with each plane is the same. Figures 34b and 34c, on the other hand, show the actual operative slip planes for niobium in tension and compression, respectively. The asymmetry associated with these latter two figures when compared to 34a shows that slip takes place more easily in the twinning sense, i.e. the ($\bar{1}$12) plane in tension and the (121) plane in compression. The twinning sense is most easily visualized by reference to Fig. 26. When the applied stress operates in the opposite direction to the twinning sense, i.e., on the (121) plane in tension and on the ($\bar{1}$12) plane in compression, glide takes place with much more difficulty. The reason for the above differences in slip mode is due to the difficulty associated with recombining the partial dislocations given in Eqs. (30) and (37) and in turn allowing them to redissociate onto a single glide plane. If the single glide plane is ($\bar{1}$12), then the resulting dislocation configuration is as shown in Fig. 35. The twin type

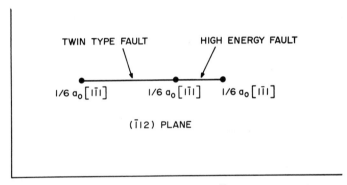

Fig. 35. Fully dissociated $\frac{1}{2}a_0[1\bar{1}1]$ dislocation on the ($\bar{1}$12) plane in a body-centered cubic crystal.

fault has already been described in Fig. 26, whereas the corresponding high energy fault can be seen in Fig. 36. The high energy of this particular fault arises from the smaller than equilibrium distance to which the nearest-neighbor atoms across the fault are positioned. This energy is even higher when the lattice possesses B2 type order, as illustrated in Fig. 36, since the nearest neighbors across the fault are all of incorrect type. Thus, the presence of the high energy fault which was argued to make glide on (121) planes relatively difficult in tension and glide on ($\bar{1}$12) planes relatively difficult in compression is expected to be even more pronounced in B2 type superlattices with large ordering energies. This seems to be the case for AgMg[79] where the calculated slip modes are shown in Fig. 37. It can be seen at once

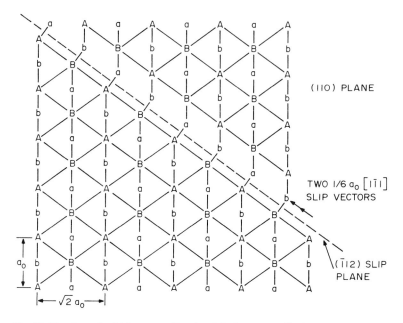

Fig. 36. Creation of a high-energy disordered fault across a $(\bar{1}12)$ slip plane in a B2 type superlattice by the passage of two partial dislocations with Burgers vector $\frac{1}{6}a_0[1\bar{1}1]$. Atom positions and bonds as in Fig. 26.

that the nontwinning slip mode is completely eliminated for all orientations in tension and restricted to a very narrow range of orientations near [001] in compression. In fact, the nontwinning slip mode in tension for AgMg is replaced by [001] slip, which is favored by both high T_C and the resolved stress. It is postulated that as T_C is increased to still higher values, as in the case of NiAl, more and more of the unit sterographic triangle is expected to correspond to [001] glide, occupying the area away from the [001] corner of the triangle. Another interesting feature of Fig. 37 is the complete absence of glide on {110} planes. A possible reason for this is related to the threefold extension of the $\frac{1}{2}a_0[111]$ dislocation in this plane as described with respect to Eq. (37). In order for such extension to occur, all three $\frac{1}{6}a_0\langle110\rangle$ partial dislocations must create disorder on their corresponding slip planes. This additional ordering energy in turn makes the dissociation of the $\frac{1}{2}a_0[111]$ dislocation more difficult, and thus its stability is reduced with respect to dissociation on {112} planes, where only two faults are created. It is thus clear from the above discussion that orientation effects are extremely important in determining the slip mode in ordered and disordered alloys.

Thus far it has been shown that the temperature dependence of the flow stress increases with decreasing temperature, either as a result of an increase

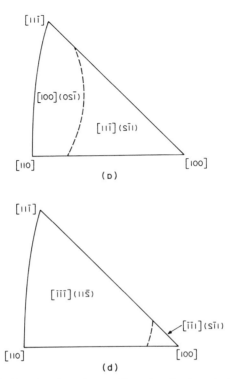

Fig. 37. Calculated orientation dependence of the active slip planes in AgMg single crystals deformed at 237.5°K: (a) tension, (b) compression. From Mitchell *et al.*[79].

in the Peierls stress or as a result of imperfect dislocation nucleation with decreasing temperature. On the other hand, in the case of $L1_2$ type superlattices, in which atomic order persists up to the melting point, the yield stress is observed to increase with increasing temperature to some maximum value, after which it continues to decrease. It is this effect that contributes greatly to the strength of superalloys[145]. This behavior can be seen by reference to Fig. 38 which shows the variation of yield stress for Ni_3Al as a function of temperature[146]. Similar results have also been obtained with Ni_3Al[147–149] as well as with Ni_3Si[150], Co_3Ti[150] and Ni_3Ga[151,152]. In all of these alloys there is essentially no temperature dependence of the yield stress at low temperatures, again indicating the small value of the Peierls stress associated with superlattice dislocations in ordered alloys. Some insight into the nature of this Peierls stress in face-centered cubic crystals may be obtained by carrying out an analysis similar to that done for the body-centered cubic lattice shown in Fig. 25. In particular, Fig. 39 illustrates two possible extensions of the $a_0[101]$ dislocation into partials. The

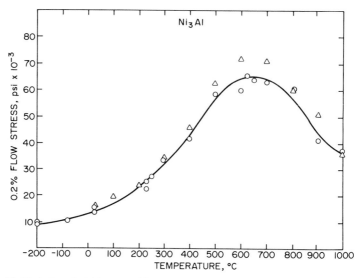

Fig. 38. Variation of yield stress with temperature for ordered polycrystalline Ni₃Al. From Davies and Stoloff[*146*]. ○, Results of Davies and Stoloff[*146*]; △, results of Flinn[*148*].

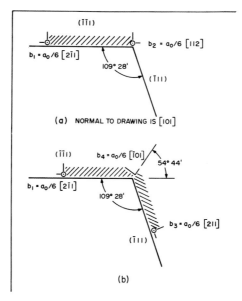

Fig. 39. Schematic illustration showing the extension of a screw type $a_0[101]$ dislocation (a) in a single plane, (b) in two planes.

shaded areas correspond to stacking faults. The dissociation shown in Fig. 39a results from the reaction[37]

$$\tfrac{1}{2}a_0[101] = \tfrac{1}{6}a_0[112] + \tfrac{1}{6}a_0[2\bar{1}1] \tag{39}$$

while that shown in Fig. 39b is the result of the dissociation given by[37]

$$\tfrac{1}{2}a_0[101] = \tfrac{1}{6}a_0[211] + \tfrac{1}{6}a_0[\bar{1}01] + \tfrac{1}{6}a_0[2\bar{1}1] \tag{40}$$

On the basis of the b^2 energy criterion for isotropic crystals, Eq. (39) involves a somewhat larger energy reduction than that given by Eq. (40). However, more-detailed calculations are required to decide between the two. Nevertheless, it is apparent that the configuration shown in Fig. 39b is sessile. In order to become glissile, it must first constrict along some critical length and then extend itself on a single slip plane such as shown in Fig. 39a. A temperature-dependent Peierls stress will thus be associated with the dislocation configuration illustrated in Fig. 39b. Atomic ordering, on the other hand, adds to the energy of the stacking faults shown in Fig. 39 since these faults now also contain APBs[11,15]. This additional APB energy causes a smaller extension of the dislocation configuration illustrated in Fig. 39b which in turn reduces the Peierls stress in a manner similar to that already described for the B2 type superlattice.

The increase in the yield stress of Ni_3Al with temperature shown in Fig. 38 was originally associated with diffusion-controlled climb of superlattice dislocations[148]. However, this hypothesis was ruled out by the fact that such processes are not important until temperatures of $T/T_M \simeq 0.5$ are attained, whereas reference to Fig. 38 shows the yield stress to be increasing even at room temperature, i.e. $T/T_M \simeq 0.2$. The reason for this behavior can be explained by a mechanism first proposed by Kear[153–155]. This is best illustrated by reference to Fig. 40 which shows the cross slip of a screw type superlattice dislocation from the $(\bar{1}11)$ to the (001) plane. The shaded regions represent APBs, while for simplicity, each of the ordinary dislocations comprising the superlattice are shown as undissociated into their corresponding partials. Cross slip is made relatively easy in this particular superlattice since no incorrect Au–Au nearest-neighbor atoms are associated with a shear type APB which lies on the (001) plane[11]. On the other hand, the cross slip process shown in Fig. 40 still requires thermal activation since each of the extended ordinary dislocations comprising the superlattice dislocation must first be coalesced into a single $a_0/2[110]$ dislocation before cross slip can occur. Such thermal energy becomes increasingly more available with increasing temperature so that cross slip becomes more frequent at elevated temperatures. Once cross slip onto (001) has occurred, however, it becomes difficult for cross slip back on to the high energy $(\bar{1}11)$ plane shown in Fig. 40. The superlattice dislocations thus become trapped on these (001)

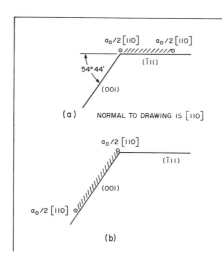

Fig. 40. Cross slip of superlattice dislocation from ($\bar{1}$11) to (001) plane in an L1$_2$ type superlattice (a) prior to cross slip, (b) after cross slip.

planes. Further glide on ($\bar{1}$11) planes can, therefore, occur only by recross slip onto these planes from (001) or by the individual glide of ordinary $a_0/2[110]$ dislocations on ($\bar{1}$11) planes with the subsequent generation of APBs. Both of these processes can account for the increase in yield stress with temperature in Ni$_3$Al and related superlattices. The question now arises as to why yield point behavior such as that shown in Fig. 38 for Ni$_3$Al occurs only in those L1$_2$ superlattices in which $T_C/T_M = 1.00$ and not in those alloys such as Cu$_3$Au, Ni$_3$Mn, and Ni$_3$Fe in which $T_C/T_M \simeq 0.50$[156]. A possible reason for this is that since atomic ordering leads to an effective increase in the stacking fault energy associated with the $\frac{1}{6}a_0\langle 112 \rangle$ type partial dislocations shown in Fig. 39[11], they became less extended with increasing ordering energy, or equivalently according to Eq. (10), with increasing T_C. Cross slip onto {100} type planes thus becomes easier in those alloys, such as Ni$_3$Al, that possess large values of T_C, even at relatively low temperatures.

The maximum in the yield stress of Ni$_3$Al shown in Fig. 38 which occurs at $T/T_M = 0.55$ and its subsequent decrease with increasing temperature may be attributed to a mechanism first proposed by Flinn[148]. In particular, in this temperature range atomic diffusion is sufficiently rapid so that the cross-slipped superlattice dislocations such as shown in Fig. 40 can glide on ($\bar{1}$11) planes while the attached APB follows along by atom migration. Alternative mechanisms have also been proposed for the maximum in the yield stress curve shown in Fig. 38 relating to the change in slip plane from

{111} to {100}[147,151] as well as to temperature-dependent changes in the elastic behavior of the alloy[149].

Figure 41 shows the entire stress–strain curve for a number of polycrystalline Ni_3Al alloys obtained in compression, at various temperatures. As in the case of the corresponding FeCo curves illustrated in Fig. 13, three well-defined stages of work hardening can be observed. It, therefore, becomes clear that the yield point, i.e., stage I, discussed thus far, represents just one

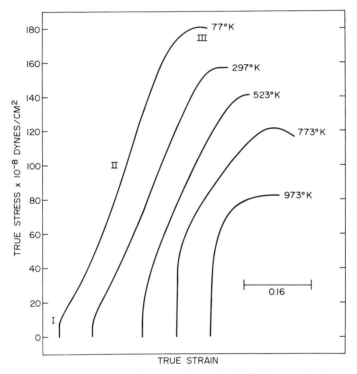

Fig. 41. Polycrystalline stress–strain curves for ordered Ni_3Al at various temperatures. From Marcinkowski and Campbell[156].

point on each of these stress–strain curves, and that the stage III flow stress shows just the reverse temperature dependence from that exhibited by the yield stress or stage I. These results can be more easily seen by reference to Fig.42. For comparison purposes, the flow stress for pure nickel in stage III, defined arbitrarily as 50% strain, is also shown as a function of temperature in Fig. 42[156]. The similarity in temperature dependence between the stage III flow stress for both Ni and Ni_3Al indicates that both processes are the same, namely a temperature-dependent Peierls stress. In Fig. 41 the stress–

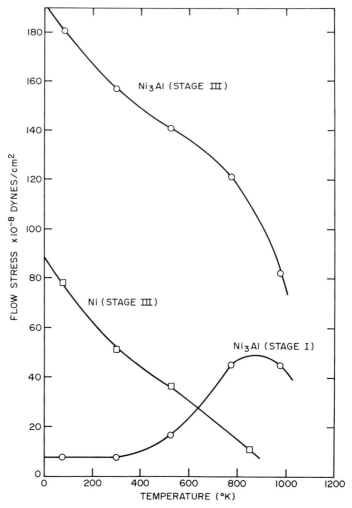

Fig. 42. Flow stress of Ni_3Al and Ni as a function of temperature for various stages of plastic deformation.

strain curves that are characterized by rapid work hardening also indicate the dangers inherent in tests such as hardness which measure a flow stress somewhere in stage II and thus are not sensitive to well-defined stages such as shown in Fig. 42[157].

Before concluding this section it is of interest to consider how the techniques of thermal activation strain rate analysis apply to ordered alloys. Methods of this type have already been applied to Cu_3Au[116] as well as to

FeCo[54]. In particular, for a fixed dislocation structure, the strain rate imposed on a single or polycrystalline material can be written as[158]

$$\dot{\varepsilon} = f(\tau^*, T) \tag{41}$$

where τ^* is that contribution to the flow stress which depends on temperature, i.e., the thermal component. More specifically, it has been shown that $\dot{\varepsilon}$ can in general be expressed in the form of an Arrhenius type equation given by

$$\dot{\varepsilon} = \dot{\varepsilon}_0 \exp\left(-\Delta H/kT\right) \tag{42}$$

where ΔH is defined as the activation enthalpy associated with the specific dislocation mechanism and $\dot{\varepsilon}_0$ is a constant. Equation (42) can also be rewritten as

$$d \ln \dot{\varepsilon}/dT = \Delta H/kT^2 \tag{43}$$

Furthermore, by use of the following relationship

$$(d \ln \dot{\varepsilon}/d\tau)_T (d\tau^*/dT)_{\dot{\varepsilon}} = -d \ln \dot{\varepsilon}/dT \tag{44}$$

Eq. (43) can be expressed as

$$\Delta H = -kT^2 (d \ln \dot{\varepsilon}/d\tau)_T (d\tau^*/dT)_{\dot{\varepsilon}} \tag{45}$$

or more compactly as

$$\Delta H = -Tv^* (d\tau^*/dT)_{\dot{\varepsilon}} \tag{46}$$

where

$$v^* = kT(d \ln \dot{\varepsilon}/d\tau)_T \tag{47}$$

The quantity v^* in the above equation is termed the activation volume and is related to the area swept out by the dislocation during its thermal activation over the particular barrier.

Figures 43 and 44 show the variation of v^*/b^3 with plastic strain for a number of temperatures for the ordered and disordered alloys of FeCo, respectively[54]. These values are all given in reduced units of b, the Burgers vector associated with ordinary dislocations. Furthermore, all values of v^* were calculated from Eq. (47) in which the term in parentheses was determined from strain-rate change measurements. It is apparent from these curves that v^* associated with stage III deformation at the lower temperatures in both ordered and disordered alloys is nearly independent of strain, and that for a given temperature, v^* for both ordered and disordered alloys possess nearly the same values. Furthermore, these values are essentially the same as those obtained for iron[54]. This suggests that in both cases the

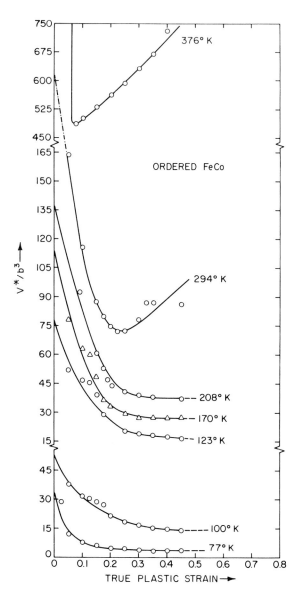

Fig. 43. Variation of activation volume with plastic strain for fully ordered FeCo. From Fong *et al.*[54].

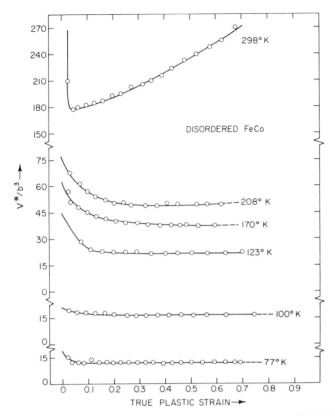

Fig. 44. Variation of activation volume with plastic strain for disordered FeCo. From Fong et al.[54].

processes are essentially the same, namely, the movement of ordinary dislocation over the Peierls barriers. On the other hand, the values of v^* associated with stages I and II in the ordered alloys as well as in the low strain region of the disordered alloy all exhibit higher values when compared with the stage III results. These larger values of v^* can all be readily attributed to the movement of coupled dislocations which is occasioned by either long- or short-range order. In both cases the area swept out by such pairs during the thermal activation process is expected to be significantly higher than that associated with a single ordinary dislocation. The increase in v^* with strain that occurs at high values of temperature and strain in both Figs. 43 and 44 is related to a reordering of the lattice at these temperatures as a result of the high density of deformation induced point defects.

Using the values of v^* obtained from Figs. 43 and 44 and the corresponding values of $(d\tau^*/dT)_{\dot{\varepsilon}}$ determined from Figs. 29 and 30, values of ΔH can be determined for both the ordered and disordered FeCo alloys as a function of plastic strain and are shown in Figs. 45 and 46, respectively. These figures again show that at high strains, i.e. stage III, ΔH is nearly the same for both ordered and disordered alloys and that these values are approximately the same as those obtained for pure iron[54]. These findings again give support to the Peierls mechanism as being responsible for the temperature dependence of the flow stress at these strains. At lower strains, i.e. stages I and II,

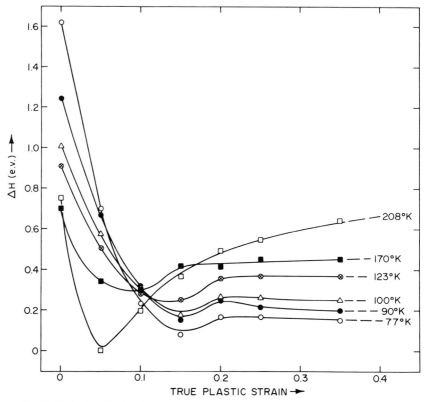

Fig. 45. Variation of activation energy with plastic strain for fully ordered FeCo. From Fong *et al.*[54].

Fig. 45 shows a rapid increase in ΔH. A somewhat smaller increase in ΔH is also noted for the disordered alloy. In both these cases, coupled dislocations play a large part in the deformation process. It would, therefore, be expected that ΔH should be small in accordance with the low Peierls stress anticipated in highly ordered alloys as discussed earlier. On the other hand, referring to Fig. 29, it has been shown that at low strains the variation of flow stress with temperature is associated with the nucleation of single or

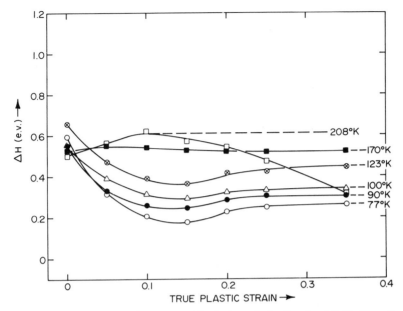

Fig. 46. Variation of activation energy with plastic strain for disordered FeCo. From Fong et al.[54].

imperfect dislocations. The same effect is expected to hold for the results shown in Fig. 30 for the disordered alloy, but to a somewhat lesser extent. The resulting values of $(d\tau^*/dT)_{\dot{\varepsilon}}$ obtained from these curves thus do not correspond to a thermal activation process, but represent an increase in the flow stress necessary to create more APBs, i.e., more disorder. The increase in flow stress with decreasing temperature thus corresponds to an increasingly larger contribution to the athermal component of stress. The resulting values of $(d\tau^*/dT)_{\dot{\varepsilon}}$ used in Eq. (45) are therefore abnormally large, in turn, resulting in abnormally high values of ΔH.

C. Work-Hardening Behavior

Perhaps the most striking effect of atomic ordering on mechanical behavior is the marked increase in the degree of work hardening that characterizes stage II[156]. This is most clearly seen in Fig. 13. Furthermore, the degree of work hardening, as defined by $\theta_{II} = d\sigma/d\varepsilon$ is essentially constant throughout all of stage II. In the case of polycrystalline Fe_3Si, θ_{II} attains a value which is about 0.3μ, while even in a single crystal oriented so as to deform on a single slip system, θ_{II} is still one-half that observed for a

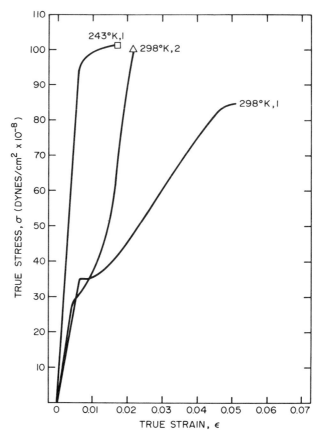

Fig. 47. Compressive stress–strain curves for single crystals of Fe₃Si obtained at 243 and 298°K.

polycrystal[34,159]. The stress–strain curve labeled 298°K, 1 in Fig. 47 illustrates the high work hardening observed in single crystals of Fe₃Si oriented for deformation via superlattice dislocations associated with a single slip system. As noted earlier, the curve labeled 243°K, 1 relates to single crystalline deformation via ordinary dislocations. If, on the other hand, the Fe₃Si crystal is oriented for simultaneous deformation on two intersecting slip systems, the curve labeled 298°K, 2 in Fig. 47 obtains. This curve shows that almost immediately after yielding, the work hardening is such as to stop further plastic deformation almost entirely, i.e., θ_{II} becomes equal to μ. In addition to the Fe₃Si and FeCo alloys, high rates of work hardening occasioned by atomic ordering have also been obtained for single[44] and polycrystalline[160,161] Fe₃Al, polycrystalline FeCO–2V[56], polycrystalline FeAl[*156*], poly[*126,156*] and single crystalline[*149*] Ni_3Al, single[*1,154,155,162–164*] and polycrystalline Cu_3Au[*116,156*], single[*128*] and polycrystalline[*58–60,135,136,156*] Ni_3Mn, single[*165,166*] and polycrystalline[*59,60,103,156,167*] Ni_3Fe, polycrystalline[*156*] Ni_3Si, polycrystalline FeRh[*156*], single crystalline[*151*] Ni_3Ga, polycrystalline CuZn[*62,156*], single[*137–139*] and polycrystalline NiAl[*139*], poly-crystalline[*141*] AuZn, single[*79,80*] and polycrystalline[*142*] AgMg, polycrystalline[*168*] CuAu, single[*120,121,169*] and polycrystalline[*82*] Mg_3Cd, and polycrystalline[*170*] Cu_3Pd. In the above observations, the NiAl, AgMg, and AuZn findings appear to be complicated by the fact that $\langle 100 \rangle$ as well as $\langle 111 \rangle$ slip occur, and that the rate of work hardening is a function of the slip direction. Also, in CuAu the work-hardening behavior is undoubtedly complicated by the internal stresses resulting from the tetra-gonality of the ordered lattice. In addition, the Mg_3Cd single crystal results are somewhat unclear concerning the work-hardening behavior of the ordered alloys with respect to the disordered ones. With the possible excep-tion of these results, on the other hand, the high work-hardening rates associated with the remaining alloys, all seem to be related to the presence of superlattice dislocations.

One of the earliest theories of work hardening in ordered alloys was formulated in terms of the intersection of superlattice dislocations on differ-ent slip systems[*171–175*]. In particular, it was postulated that intersecting dislocations would give rise to jogs, which in turn would produce APB tubes. Following the treatment given in Vidoz and Brown[*171*], Fig. 48

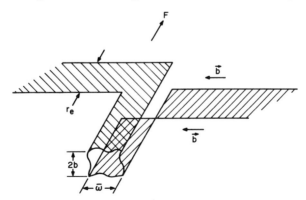

Fig. 48. Jog on a superlattice dislocation.

shows a screw type superlattice dislocation lying on the primary slip plane which has been jogged by a second screw type forest superlattice dislocation (not shown). As the superlattice dislocation moves rigidly in the direction F it leaves behind an APB tube, the top and bottom faces of which consist of an APB of width $\bar{\omega}$, while the two side faces can be thought of as an edge dislocation dipole of separation $2b$. The quantity $\bar{\omega}$ is dependent upon the relative velocities of the primary and forest dislocations. More specifically, it is related to the equilibrium extension of the superlattice r_e, given by Eq. (7), according to

$$\bar{\omega} = \lambda r_e \tag{48}$$

where λ is a constant whose value depends on the mutual velocities of the intersecting dislocations. When a length L of the superlattice dislocation shown in Fig. 48 moves forward rigidly a distance s, the energy expended is given by

$$E_{TE} = sLn2(E_{TO} + E_{TD}) \tag{49}$$

where n is the number of APB tubes per unit length, while E_{TD} is the energy per unit length of one edge dislocation dipole, and E_{TO} is the APB energy per unit length of one of the APB tube faces, and is given by

$$E_{TO} = \bar{\omega}\gamma_0 \tag{50}$$

The energy expended in Eq. (49) must be supplied by the work of the applied stress $2\tau bLs$ which gives

$$\tau = (n/b)(E_{TO} + E_{TD}) \tag{51}$$

If now the alloy is disordered without changing the jog density, an equation analogous to that given by Eq. (51) can be found to be

$$\tau' = (n/b)E_{TD} \tag{52}$$

The difference between Eqs. (51) and (52) thus represents the increase in flow stress due to atomic ordering and can be written as

$$\Delta\tau = (n/b)E_{TO} = (n\lambda/2\pi)\mu b = (n'\lambda/8\pi)\mu b \tag{53}$$

where n' corresponds to the jog density, which is related to the APB tube density by $n' = 4n$. Furthermore, the jog density n' can be assumed proportional to $\rho_F^{1/2}$ where ρ_F is the forest dislocation density[173]. An analysis similar to that given above can be carried out for jogged edge dislocations. In this case the jogs move conservatively but leave behind tubes whose four faces are comprised of APBs. In fact, it has been argued in Schoeck[172] that it is the jogged edge dislocations that are responsible for stage II work hardening in superlattices.

The results obtained from both deformation and transmission electron microscopy studies of single crystals of Fe_3Si[34,159], Fe_3Al[26,38,44], and Cu_3Au[163] oriented for slip on a single slip system are incompatible with the jog theory discussed above. In all cases, strong stage II work hardening was observed, while no evidence for APB tubes was detected. Furthermore, the longer lengths of the screw segments of the dislocations compared to those of the edges, such as can be seen for the case of Fe_3Si in Fig. 4, rule out the hypothesis that the edge dislocation segments are jogged more heavily than the screws. The above experimental observations are also in disagreement with the hypothesis that dislocation–APB intersections decrease the effective spacing between APB[148] and thus give a strengthening similar to that predicted by Eq. (33), since that model also depends on the operation of intersecting slip systems.

A third model of stage II work hardening has been proposed; it depends upon the trapping of screw dislocation segments on low energy {100} planes as illustrated in Fig. 40[153–155]. The difficulty with this theory is that it has strict applicability only to $L1_2$ superlattices. Furthermore, there seems to be no obvious way in which to relate an increase in trapping to increased strain.

Because of the above difficulties it has been postulated that the large value of θ_{II} associated with ordered alloys oriented for single slip was associated with the formation of dislocation dipoles[26,34,38,44,47,50,159,176,177] or more generally, multipoles[163]. Furthermore, the breaking of such dipoles was shown to be an athermal process[179]. The above process is thus seen to be based on the Taylor theory of work hardening[37]. For screw dislocations, the locking or maximum passing stress is readily found to be[47]

$$\tau_T = \mu b/4\pi y \qquad (54)$$

where y is the spacing between parallel slip planes. For edge dislocations, 4π in the above equation is replaced by $8\pi(1 - \nu)$[47]. In order to obtain τ_T as a function of plastic strain, it can be assumed that the strain and dislocation density are related as follows:

$$d\Gamma = d\rho bL \qquad (55)$$

where L is the distance moved by each dislocation. This may be related to ρ as

$$L = \rho^{-1/2} \qquad (56)$$

Combining Eqs. (55) and (56) and integrating gives

$$\Gamma = 2b/L \qquad (57)$$

Realizing that $L \equiv y$ in Eq. (54) and combining this equation with (57) leads to

$$\tau_T = (\mu/8\pi)\Gamma \qquad (58)$$

Thus, the flow stress is linearly related to Γ, in agreement with the observed results, such as shown in Figs. 13 and 47. On the other hand, Taylor's theory, presented in Ref. [37], assumed L to be constant and obtained the well-known parabolic relation for flow stress given by

$$\tau_T = \alpha\mu(\Gamma b/L)^{1/2} \tag{59}$$

where α is a constant on the order of 0.1. On the other hand, in deriving the above relationship, Eq. (56), along with the relation given by

$$\Gamma = \rho b L \tag{60}$$

was employed, which when combined, gives

$$\Gamma = b/L \tag{61}$$

It is apparent that Eqs. (61) and (62) are equivalent to Eqs. (57) and (55), respectively. Thus, according to Eq. (61), L in Eq. (59) cannot be constant. Substitution of Eq. (61) into (59) thus gives

$$\tau_T = \alpha\mu\Gamma \tag{62}$$

which is in agreement with the formulation given by Eq. (58).

The concept of cross slip in ordered alloys is a particularly important one. Six important types of cross slip can be seen by reference to Fig. 49. In all cases the dislocations are straight screw types which run normal to the drawing. In particular, Fig. 49a shows the cross slip of an ordinary dislocation around some suitable obstacle under the action of the applied stress τ. Figure 49a', on the other hand, shows a pair of dislocations of opposite sign cross slipping under the influence of their mutual stress fields, as well as the applied stress, with subsequent self-annihilation. Figure 49b illustrates the manner in which a blocked superlattice dislocation is able to cross slip from one slip plane to another under the action of the applied stress τ. This process has been treated in detail in Marcinkowski et al.[178]. Unlike the cases illustrated in Figs. 49a and 49a', an energy barrier generally exists for the cross slip of superlattice dislocations, and is occasioned by the APB energy. The physical argument for this is that during the very initial stages of cross slip, the separation between the two ordinary dislocations comprising the superlattice dislocation is virtually unchanged from that on the primary slip plane, whereas APB contributions are now present on both the primary as well as the cross slip planes. Under a suitably large stress τ_c, however, the process can be made spontaneous. This stress is given by

$$\tau_c = \frac{\gamma_2 - \gamma_1 \cos\alpha}{b[\cos\beta \cos\alpha + \cos(\beta + \alpha)]} \tag{63}$$

where γ_2 and γ_1 are the APB energies on the cross slip and primary planes, respectively, while α is as defined in Fig. 49b and β is a parameter relating to the resolution of τ on the slip plane. Specifically, when $\beta = 0$, the maximum

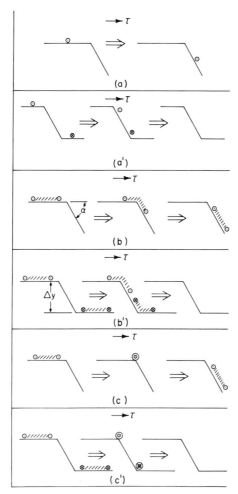

Fig. 49. Cross slip of (a), (a′) ordinary dislocations, (b), (b′) extended superlattice dislocations, (c), (c′) constricted superlattice dislocations by (a), (b), (c) externally applied stress by (a′), (b′), (c′) internal stress from second dislocation, in addition to externally applied stress.

value of τ_c occurs on the primary plane and is a minimum on the cross slip plane, while for $\beta = -\pi/2$, it is zero on the primary plane and a maximum on the cross slip plane. Inspection of Eq. (63) shows that for $\gamma_2 \simeq \gamma_1$, cross slip will always be difficult since α is generally much less than zero. On the other hand, for $\gamma_1 > \gamma_2$, as in the case of the $L1_2$ superlattices discussed earlier in connection with Fig. 40, it is possible that cross slip could occur spontaneously in the absence of an externally applied stress.

Figure 49b′ illustrates the mutual cross slip of a pair of superlattice dislocations of opposite sign with subsequent mutual annihilation. This case has
also been treated in detail[179] and it has been shown that τ_c given by
Eq. (63) is reduced considerably due to the internal attractive stress of the
second dislocation. This can be seen from the numerical results shown in

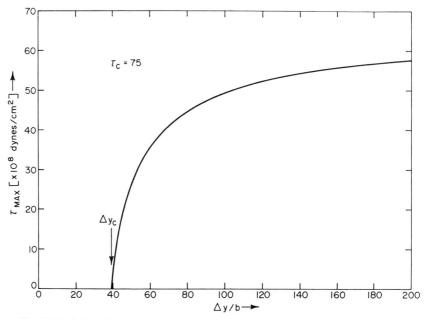

Fig. 50. Relationship between the applied stress and the initial vertical separation for
spontaneous cross slip of a pair of superlattice screw dislocations of opposite sign.

Fig.50 which pertains to the fully ordered FeCo alloy. For $\Delta y \to \infty$, i.e., the
case shown in Fig. 49b, $\tau_{max} \to \tau_c = 75 \times 10^8$ dyn/cm^2 which is given by
Eq. (63). However, when Δy, as defined in Fig. 49b′, reaches the value of 40b,
$\tau_{max} = 0$; i.e., cross slip occurs spontaneously in the absence of an externally
applied stress.

It was initially argued[180] that in order for cross slip of a superlattice dislocation to occur, it would first have to be constricted as illustrated in Fig. 49c,
much in the same way as that required for the $\frac{1}{6}a_0 \langle 112 \rangle$ partials in a face-
centered cubic crystal. However, in the latter case, such constriction is a geo-
metric necessity to eliminate the edge components associated with the two
partial dislocations, and thus need not occur for a superlattice dislocation. It
has been argued[181], however, that for sufficiently large values of γ_0, the extension of the superlattice would be very small, in accordance with Eq. (7),
so that cross slip in the constricted manner shown in Fig. 49c would be

favored. To find the critical conditions under which this occurs, it can be assumed that the trailing dislocation of the pair would have to be driven to within a distance b of the leading dislocation before the superlattice dislocation could be considered as coalesced. The total force on the trailing dislocation would then be

$$\tau_c b + \gamma_1 = \mu b^2 / 2\pi b \qquad (64)$$

The term on the right-hand side of the equal sign in the above equation is simply the repulsive force between the dislocation pair at a distance b. It is assumed that $\beta = 0$ in Eq. (64); i.e., the maximum stress acts on the primary slip plane. Assuming this also to be true for Eq. (63), it is possible to equate τ_c from both equations to give the critical conditions for superlattice dislocation coalescence. In particular

$$\frac{\gamma_2 - \gamma_1 \cos \alpha}{2b \cos \alpha} \geq \frac{\mu}{2\pi} - \frac{\gamma_1}{b} \qquad (65)$$

For simplicity, consider the case where cross slip is confined to planes of the same type, i.e., $\gamma_1 = \gamma_2$. For $\{111\}$ type planes in body-centered cubic crystals, $\alpha = 60°$, while for $\{111\}$ planes in face-centered cubic crystals, $\alpha = 70°32'$. Under these conditions, $\gamma_1 > 1000$ ergs/cm^2 for most materials. Reference to Table II shows that superlattice dislocation coalescence prior to cross slip will never occur since the highest value of $\gamma_0 = 300$ ergs/cm^2 and occurs for Ni$_3$Al. For the case shown in Fig. 49c', coalescence may be aided by the internal stress field of the second dislocation, so that the conditions given by Eq. (65) could be relaxed somewhat.

Returning again to the case of cross slipping ordinary dislocations in Figs. 49a and 49a', it is apparent that cross slip in the former case is spontaneous under a vanishingly small applied stress τ. In the latter case, however, cross slip occurs spontaneously in the absence of an applied stress due to the internal stress field of the second dislocation. When there is a lattice friction force, $F_f = \tau_f b$, present in Fig. 49a, τ_c is readily found to be given by

$$\tau_c = F_f / (b \cos \alpha) \qquad (66a)$$

where β is assumed to be zero. If, on the other hand, the ordinary dislocation moves in an ordered alloy, a ribbon of APB is left behind and F_f in Eq. (66a) is replaced by the APB energy γ_0 so that (66b) becomes

$$\tau_c = \gamma_0 / (b \cos \alpha) \qquad (66b)$$

Similarly, for the case shown in Fig. 49a' where lattice friction is included

$$\tau_c = \frac{F_f}{b \cos \alpha} - \frac{\mu b}{2\pi r \cos \alpha} \qquad (67)$$

where r is the separation of the dislocations just prior to mutual cross slip. Again, if atomic ordering is present F_f is replaced by γ_0.

It now becomes possible to discuss the role of cross slip in the work-hardening behavior of ordered alloys. Again, consider the expression for the change in Γ associated with an increase in ρ. The transmission electron micrograph in Fig. 4 shows that even in stage II a large number of screw dislocations are undergoing self-annihilation as can be seen in the regions labeled C. Furthermore, the annihilation is occurring between screw type superlattice dislocations, so that the mechanism is most likely that shown in Fig. 49b'. A simple way in which to include this annihilation factor into Eq. (55) would be to write it as

$$d\Gamma = (1/\delta)\, d\rho bL \tag{68}$$

where δ is a parameter that varies from 1 for no dislocation annihilation to 0 for complete dislocation annihilation. In this latter case

$$d\rho/d\Gamma = 0 \tag{69}$$

Physically, this means that with each increment of strain, the number of dislocations generated is balanced by an equivalent number that are removed by self-annihilation, i.e., $d\rho = 0$ or $\rho = C$. Combining Eqs. (69), (54), and (56) leads to the following flow stress associated with stage III deformation:

$$\tau_T = (\mu b/4\pi)C^{1/2} \tag{70}$$

Furthermore, Eq. (58) for the flow stress in stage II, must be modified to incorporate the effect of dislocation annihilation given by Eq. (68). In particular, it now becomes

$$\tau_T = (\mu\delta/8\pi)\Gamma \tag{71}$$

Equation (70) unfortunately does not, by itself, predict the stress level at which stage III occurs. On the other hand, it has been previously stated that the deformation taking place in this stage is principally through the aid of ordinary dislocations. The stress to move these ordinary dislocations will thus be given by

$$\tau_{(III)} = \gamma_0/b \tag{72}$$

The ordinary dislocations will also be able to cross slip quite easily in stage III since γ_0 no longer acts to keep the ordinary dislocations confined to a single slip plane by means of the superlattice dislocation. Another important question at this juncture concerns the reason for the stability of ordinary dislocations with respect to superlattice dislocations in stage III. The most obvious reason for this lies in the instability of the superlattice dislocation dipoles at these stage III stress levels as described in connection with Fig. 11. All of the observations are in accordance with the characteristic wavy nature of the slip plane traces commencing as soon as stage III is attained as observed in ordered FeCo[55], Ni$_3$Al[126], Fe$_3$Al[26,38,161], and Ni$_3$Mn[136]. Nearly horizontal wavy slip plane produced APB traces typical of stage III deformation, can be seen by reference to the transmission electron micrograph of Fig. 51 obtained from a Fe$_3$Al alloy possessing B2 type order[78]. The oval boundaries are thermally produced APBs.

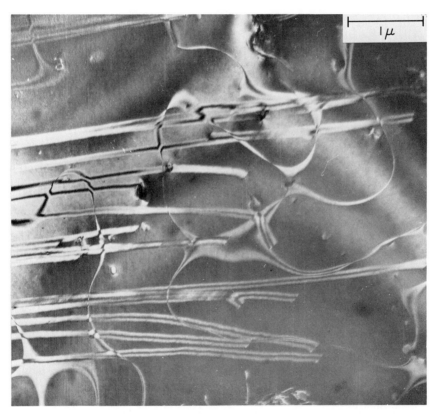

Fig. 51. Transmission electron micrograph showing the production of nearest neighbor antiphase boundaries by the irregular motion of screw dislocations in the Fe$_3$Al superlattice. From Marcinkowski and Brown[78].

By subtracting out the temperature-dependent contribution to the flow stress associated with both the ordered and disordered FeCo alloys shown in Figs. 13 and 14, respectively, the athermal stress–strain curves can be obtained[54], as shown in Fig. 52. It is immediately clear by inspection of

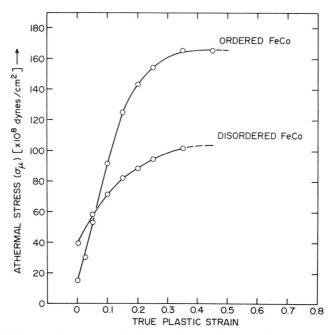

Fig. 52. Stress–strain curve for FeCo in which only the athermal component of stress is shown. From Fong *et al.*[54].

these curves that the athermal contributions to the flow stress provide the major contributions to the overall flow stress and thus determine the gross features of the stress–strain curve. Thus, stage II work hardening is essentially athermal, in agreement with the Taylor theory discussed earlier. The same holds true for stage III, so that the cross slip processes which characterize this stage can also be assumed to be athermal. The temperature dependence in stage III is related to the Peierls stress, while that in stage II is related to temperature-dependent ordinary dislocation generation along with perhaps a weak Peierls stress contribution as discussed earlier.

It is now possible to evaluate the stage III flow stress determined from Eq. (72) and compare it with some observed results. Considering both the ordered FeCo and Fe_3Si alloys, and using the data given in Table II, τ is found to be 63.5×10^8 dyn/cm^2 and 50.2×10^8 dyn/cm^2. Multiplying both of these values by the appropriate orientation factor, i.e., 2, to convert shear stress to normal stress, gives values for the stage III flow stress which are in reasonable agreement with the experimentally observed values of 160×10^8 dyn/cm^2 in Fig. 52 and 110×10^8 dyn/cm^2 obtained from reference 34, respectively.

Since γ_0 varies with the long-range order parameter S as S^2, as can be seen from Eq. (31), $\tau_{(III)}$ should decrease with decreasing S. This is vividly illustrated for the series of room temperature compressive stress–strain curves shown in Fig. 53, corresponding to various degrees of long-range order in FeCo[55]. It is interesting to note that the above findings are just the reverse of that obtained for the stage III flow stress in face-centered cubic metals and disordered alloys. In these materials $\tau_{(III)}$ increases with decreasing stacking

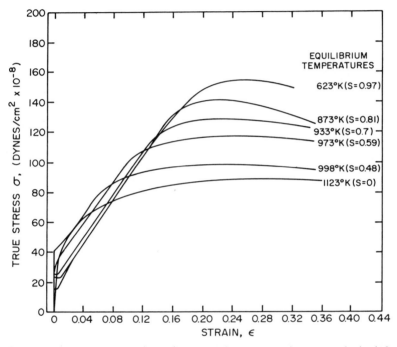

Fig. 53. Room temperature (297°K) compressive stress–strain curves obtained from polycrystalline FeCo alloys possessing various degrees of long-range order S. From Marcinkowski and Chessin[55].

fault energy[180,182]. This would also be the case for ordered alloys if the superlattice coalescence model shown in Figs. 49c and 49c′ were operative, since the stage III flow stress would then be given by Eq. (64), which is similar to expression given in Seeger[182].

The stage II flow stress given by Eq. (71), which is based upon dipole locking, holds only when a single slip system is operating, as in the case of the stress–strain curve labeled 298°K, 1 in Fig. 47 for the Fe$_3$Si single crystal. Since it has been shown that polycrystals of Fe$_3$Si give values of θ_{II} twice that of corresponding single crystals oriented for single slip, it follows that dislocation intersection must make an important contribution to θ_{II}. There seem to be two mechanisms by which a dislocation lying on a primary slip plane can pass through a forest of intersecting dislocations. On the one hand, the dislocations can be regarded as perfectly flexible, in which case they bow out between the forest dislocations according to the Orowan mechanism. The stress to accomplish this is given by Eq. (25) with γ_0 assumed to be zero. Equating R in this equation with L in Eq. (57), the flow stress for Orowan strengthening in stage II can be written as

$$\tau_O = \frac{2m-1}{2(m-1)} \frac{\mu}{8\pi} \Gamma \left[\ln \left(\frac{8b}{\varepsilon\Gamma} \right) - 1 \right] \tag{73}$$

Since the logarithmic term varies rather slowly with Γ, τ_O can be taken as proportional to Γ, again leading to a relation similar to that given by the dipole model which led to Eq. (71). The above model appears to be in good agreement with the transmission electron microscopy observations in Fig. 54 which shows a fully ordered Fe_3Si oriented for equal amounts of slip on two equivalent intersecting slip systems[34]. The plane of the foil is $(\bar{1}01)$, and the dislocations lying therein have Burgers vectors that lie along [111], i.e., the first slip system, while the second equivalent slip system consists of a second (011) plane rotated by 60° about the $[1\bar{1}1]$ axis. The dislocations lying in this second plane have Burgers vectors which lie along $[\bar{1}\bar{1}1]$, and in those cases where they are of pure screw type, their projections lie along $[0\bar{2}0]$ of the foil plane. The dislocation loops lying in the plane of the foil in Fig. 54 are seen simply to bend around the forest dislocations in the Orowan sense. Strictly speaking, Eq. (73) holds only for ordinary dislocations, so that the more exact expressions given by Eq. (23) should be used for superlattice dislocations. However, even these are good only for large values of R, and for still better approximations, numerical results such as shown in Fig. 7 must be employed. From this figure it will be noted that as R decreases below some critical value, the superlattice dislocation is unable to bow out as a coupled pair. This presents a mode of superlattice dislocations uncoupling in the Orowan approximation in much the same way as dipole uncoupling was the mode in the Taylor approximation.

If the superlattice dislocations move through the forest in a rigid manner, then the flow stress given by Eq. (51) should be employed. Combining this equation with Eq. (57) gives

$$\tau = (\Gamma/8b^2)(E_{TO} + E_{TD}) \tag{74}$$

so that once again τ is a linear function of Γ. It is apparent from the Fe_3Si results[34] that the dipole locking and forest cutting contributions to the flow stress are additive. In particular, the contributions from Eqs. (74) and (71) could be added together directly if both these methods were operative since both are based upon infinite straight dislocations. On the other hand,

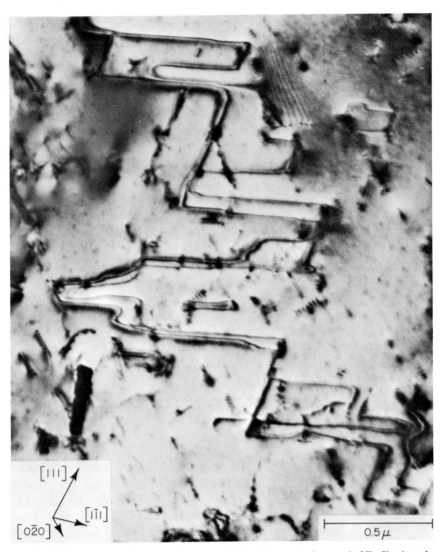

Fig. 54. Transmission electron micrograph obtained from a single crystal of Fe₃Si oriented for double slip after being compressed to a plastic strain of 0.007. Normal to the foil is [Ī01]. From Lakso and Marcinkowski[*34*].

the Taylor model for curved dislocations is much more complex and would have to be solved numerically for superlattice dislocations. Results of this type, however, have been performed for ordinary dislocations, where it has been shown that the curvature of the dislocation line greatly increases the passing stress of unlike dislocations as the radius of curvature decreases with respect to the vertical passing distance[183].

Before concluding the present section on work-hardening behavior in ordered alloys, it is interesting to point out the diversity that may arise in the shape of the stress–strain curves associated with the more complex superlattice types. In particular, Fig. 55 shows a schematic illustration of the stress–

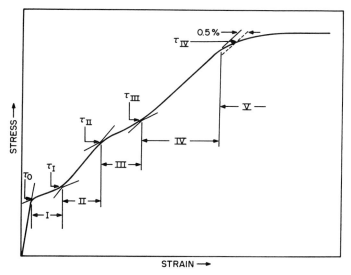

Fig. 55. Schematic illustration of a stress–strain curve obtained from superlattice alloys in the vicinity of Fe₃Al. From Leamy et al.[44].

strain curve for single crystals of Fe_3Al[55]. The curve is seen to consist of five stages of deformation. Stages I and II are associated with the motion of perfect DO_3 type superlattice dislocations of the type shown in Table I, while stages III and IV correspond to the motion of imperfect DO_3 type superlattice dislocations that generate NNN APBs. Finally, stage V is characterized by the motion of still less-perfect DO_3 type superlattice dislocations that generate NN APBs. These latter dislocations are of the same type that would be present in the corresponding disordered alloy. The above behavior is brought about by the uncoupling of the superlattice dislocation at its weakest APB when the force on the dislocation from the applied stress just matches or exceeds the APB tension, i.e., when the condition given by Eq. (72) is attained.

D. Influence of Grain Size

It has been shown repeatedly[184] that the flow stress varies with grain size D for a wide variety of materials according to the following relation, i.e. the Hall–Petch relation:

$$\sigma = \sigma_0 + kD^{-1/2} \tag{75}$$

where σ_0 and k are material constants. From measurements carried out with ordered alloys of FeCo[185], FeCo–2V[187,188], Ni$_3$Mn[115,186], and Ni$_3$Fe[167], it has been shown that atomic ordering leads to marked increases in k. This can be seen by reference to Fig. 56 which shows the

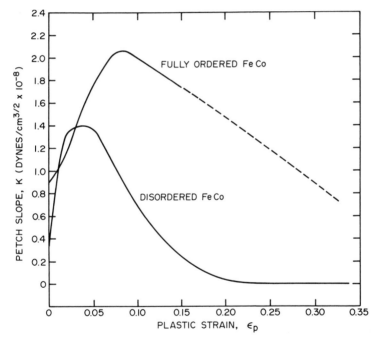

Fig. 56. Variation of the Hall–Petch slope K with plastic strain for both fully ordered and disordered FeCo. From Marcinkowski and Fisher[185].

variation of k with plastic strain for both ordered and disordered FeCo. In addition to k being generally higher for the ordered alloy for most strains, k is also seen to pass through a maximum as a function of plastic strain. The detailed plots of σ versus $D^{-1/2}$ for the ordered alloys, from which the k values shown in Fig. 56 were obtained, can be seen in Fig. 57.

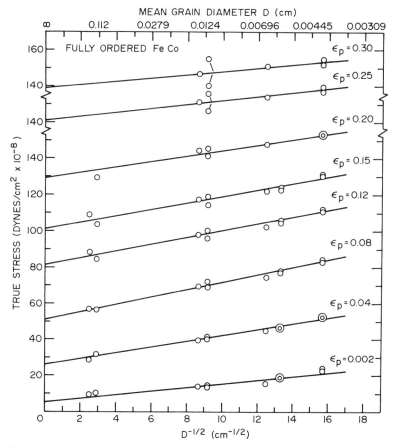

Fig. 57. Variation of the flow stress with the reciprocal of the square root of the grain diameter in a fully ordered FeCo alloy at various plastic strains ε_p. From Marcinkowski and Fisher[185].

In order to understand the physical significance of Eq. (75), consider the schematic illustration shown in Fig. 58. Slip is activated in grain number 1 by an applied stress σ which produces a shear stress τ on the slip system in this grain. The values of σ and τ are connected by some mean orientation factor $\bar{m}$, commonly called the Schmid factor, in the following manner:

$$\sigma = \bar{m}\tau \tag{76}$$

The value of $\bar{m}$ depends upon the crystal structure, but is generally taken as 2. It has been shown[37] that the glide within grain 1, as shown in Fig. 58, acts as a freely slipping crack of length D, and produces a stress concentration at some point r within grain 2 of

$$\tau(D/r)^{1/2} \tag{77}$$

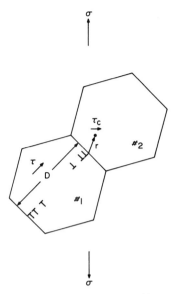

Fig. 58. Stress concentration effect produced by a grain boundary.

If, on the other hand, there is a frictional resistance to the motion of the dislocations in grain 1 given by τ_f, then τ in Eq. (77) must be replaced by $\tau - \tau_f$. Assuming now that a dislocation source exists at r which requires a stress τ_c to become activated, it follows that

$$\tau - \tau_f = \bar{m}\tau_c \tag{78}$$

The orientation factor $\bar{m}$ must again be introduced in the above expression since the slip systems in grains 1 and 2 are in general not coincident. Combining Eqs. (76), (77), and (78) leads to Eq. (75) where $\sigma_0 = \sigma_f$ and $k = \bar{m}^2 r^{1/2}\tau_c$. The generally higher value of k for ordered alloys compared with the corresponding disordered alloys can now be associated with the higher values of τ_c required to nucleate superlattice dislocations as compared to ordinary dislocations. In addition, because of the restrictive motion of superlattice dislocations compared to ordinary dislocations, $\bar{m}$ may also be higher for ordered alloys.

Inspection of Fig. 56 shows that k increases rather rapidly with strain in the very early stages of plastic deformation when work hardening is highest. It was originally thought[185] that this behavior was associated with the increase in τ_c occasioned by the increased internal stresses given by Eq. (54). It seems more reasonable, however, to associate this Taylor contribution to the flow stress with the σ_0 term in Eq. (75). The increase in k with strain may then be related to the very nature of the dislocation source itself. In particular, with increasing strain, more and more ordinary dislocations are generated in an ordered lattice, thus raising τ_c. Finally, when stage III is attained, σ_0 is raised to a level corresponding to the APB tension. At this stress level, ordinary dislocations can be nucleated in an ordered lattice with the same stress concentration, i.e. τ_c, as that required for ordinary dislocations in a disordered lattice. The above arguments also hold for the disordered FeCo alloy, which actually consists of a very strong degree of short-range order. Since, as has been noted earlier, more and more ordinary dislocations are nucleated in stages I and II as the temperature is lowered, k should be expected to increase with decreasing temperature, which is in fact observed[187,188]. As a result of the above discussions it seems appropriate to express the flow stress associated with polycrystalline deformation as[158]

$$\tau = \tau^* + \tau + kD^{-1/2} \tag{79}$$

where τ^* and τ are the thermal and athermal components of the flow stress, respectively, and $kD^{-1/2}$ is the grain size effect.

E. Composition Effects

The superlattices treated thus far have all been assumed to be of the stoichiometric type, i.e. AB or AB_3. Considering, for example, the AB type superlattice, it is intuitively obvious that the further the alloy deviates from this composition, the smaller will be the APB energy since there will be a smaller number of AB atom pairs capable of undergoing the disordering reaction given by Eq. (9) when the APB is created. In particular, for the specific case of an APB lying on the {110} plane in the B2 type superlattice[189]

$$\gamma_{(110)} = \frac{16vS^2n^2}{a_0^2\sqrt{2}}, \qquad n \le 0.5 \tag{80}$$

where n refers to the atom fraction of solute atoms. For the equiatomic alloy in which $n = 0.5$, Eq. (80) reduces to that given by (31). More general equations for γ_0 have been given in Leamy et al.[190].

It follows from the above analysis that decreasing n in Eq. (80) should be equivalent to decreasing the degree of long-range order for a given alloy composition. This is vividly demonstrated in Fig. 59 which shows the stress–strain curves corresponding to two fully ordered FeCo alloys, i.e. $S = 1$ with Co concentrations corresponding to 25 and 70 at. %[189]. It will be noted first of all that the yield stresses, i.e. stage I, for these nonstoichiometric alloys is generally higher than that for the equiatomic alloys given in Fig. 13.

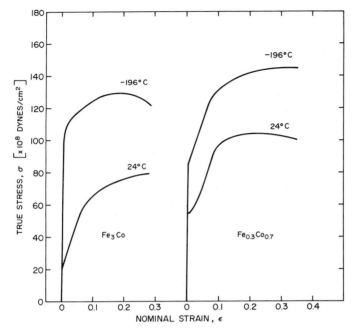

Fig. 59. Effect of composition on the compressive stress–strain curves of fully ordered polycrystalline FeCo alloys.

This observation is in accord with the earlier hypothesis discussed in connection with the yield stress–temperature data of Fig. 15 in that as the degree of atomic order decreases, the ordinary dislocations become increasingly more independent of one another, in preference to existing exclusively as superlattice dislocations. These ordinary dislocations, upon nucleation, leave ribbons of APB in their wake and thus raise the yield stress. Decreasing temperature has an even greater effect in promoting the nucleation of ordinary dislocations for nonstoichiometric alloys since the relatively weak APB tension in these alloys means that two nearly independent ordinary dislocations must be nucleated in sequence. The large thermal energy fluctuations required to accomplish this are minimal at low temperatures.

A second interesting feature of the stress–strain curves in Fig. 59 is the flow stress associated with stage III. It will be recalled that stage III has been postulated to arise from the nearly complete uncoupling of all superlattice dislocations. Since the flow stress in this stage has been predicted as that given by Eq. (72), $\tau_{(III)}$ should be expected, from Eq. (80), to increase successively from Fe_3Co to $Fe_{0.3}Co_{0.7}$ to FeCo for a given temperature. Reference to Figs. 59 and 13 reveals that this is indeed the case. A final inspection of Fig. 59 shows that although well-defined linear stage II work-hardening stages are present, they are restricted in length due both to the high stage I yield stresses as well as the low stage III flow stresses.

Results similar to those discussed above have been obtained for non-stoichiometric Fe₃Si alloys[191] as well as nonstoichiometric Fe₃Al alloys[38,44,161,192]. A minimum in the yield stress at the equiatomic composition has also been observed for AgMg alloys[193]. It has also been demonstrated that heavily cold-worked iron–nickel alloys exhibit their greatest changes in electrical resistivity at the stoichiometric Ni₃Fe composition[127] again indicating the greater destruction of long-range order via stage III deformation at these compositions. Additional verification of the above concepts can be obtained by reference to the tensile test data shown in Fig. 60 for FeAl alloys[38,194]. The large drop in both the elastic limit as well as the yield stress in the vicinity of Fe₃Al is related to the changeover in the mode of plastic deformation from ordinary to superlattice dislocations. A similar large drop in the tensile strength is not noted near

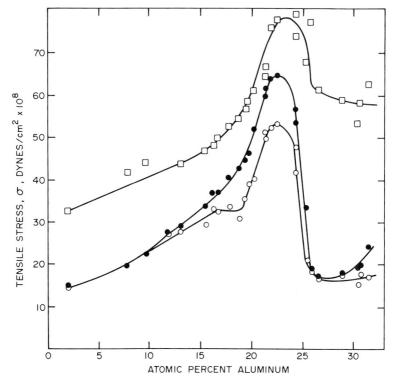

Fig. 60. Variation of the indicated mechanical properties with composition for fully ordered polycrystalline FeAl alloys (477°K). From Leamy[38]. ○ Elastic limit; ● 0.2% offset stress; □ tensile strength.

Fe_3Al, and is related to the fact that the deformation mode corresponding to these high strains takes place very nearly in stage III, so that the flow stress is governed by an equation of the type given in Eq. (72).

IV. Relationship between Twinning, Fracture, and Atomic Ordering

Twin and crack lamellae are somewhat similar[114] in that they can be described in terms of dislocation arrays such as illustrated in Fig. 61. In the case of the shear and tensile crack lamellae the dislocations are termed crack dislocations[195], whereas in the case of the twin lamellae, the dislocations are termed twin or partial dislocations[196]. All of the lamellae shown in Fig. 61, whether they consist of infinite straight dislocations or coaxial circular dislocations, become stable in the presence of the externally applied stress τ_A. The total energy of the crack lamellae may be written as

$$E_T = \sum E_S + \sum E_I + E_{\gamma_c} - E_\tau \qquad (81)$$

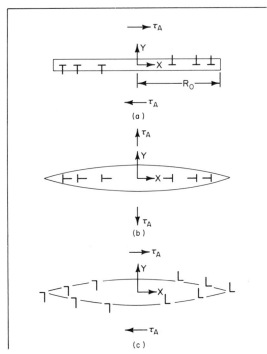

Fig. 61. Schematic illustration of (a) shear crack lamella, (b) tensile crack lamella, (c) twin lamella. From Fong *et al.*[114].

where, as in Eq. (1), $\sum E_S$ represents the self-energy of all the N dislocations comprising the crack lamella, while $\sum E_I$ is the interaction energy between all of the dislocations within the lamella. The E_{γ_c} term in Eq. (81) is associated with the surface energy of the two free surfaces comprising the crack and for a circular or penny-shaped crack is given by

$$E_{\gamma_c} = 2\gamma_c \pi R_0^2 \tag{82a}$$

where R_0 is the radius of the crack and γ_c is the energy per unit area associated with a single surface. The last term in Eq. (81) is related to the energy contribution from the applied stress τ_A and for a circular crack is given by

$$E_\tau = \pi \sum_{i=1}^{N} \tau_A b^c R_i^2 \tag{82b}$$

where b^c is the Burgers vector associated with a single crack dislocation and R_i is the radius of the ith crack dislocation.

Equation (81) also holds for a twin lamella except that now E_S and E_I represent the self- and interaction energies associated with the partial glide dislocation loops shown in Fig. 61c. On the other hand, E_{γ_T} must now be replaced by

$$E_{\gamma_T} = \alpha_T \pi R_0^2 \tag{83}$$

where γ_T is the stacking fault energy that may be associated with the outermost dislocation loop comprising the twin lamella in Fig. 61c[196]. Finally, the E_τ term in Eq. (81) is similar to that for a crack, but now b^c must be replaced by b^T, the Burgers vector associated with a twinning dislocation. A Griffith type condition given by $\partial E_T/\partial R_0 = 0$ and $\partial^2 E_T/\partial R_0^2 < 0$ can also be associated with both the crack lamellae as well as the twin lamella shown in Fig. 61[114]. It is interesting to note that the Griffith criterion simply corresponds to the maximum in energy–distance plots of the type shown in Fig. 6.

The effect of atomic ordering on the twinning mode of deformation is profound. In particular, it was first predicted by Laves[197] that atomic ordering would make twinning very difficult if not impossible. The reason for this can be most readily seen by referring to Fig. 62 which illustrates the (110) projection of a B2 type superlattice. That portion of the crystal lying above the dashed line which lies parallel to the ($\bar{1}$12) plane is in a "twin" relationship to that lying below this line. This is most easily seen by referring to the unit cells, which are shown dotted, in both regions of the crystal. Strictly speaking, the structure above the dashed line in Fig. 62 is not twinned since the unit cell in this region is different from that shown in the

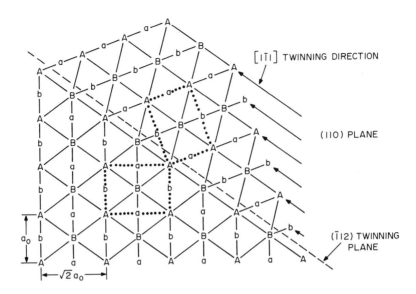

Fig. 62. Creation of incorrect atom pairs in a B2 type superlattice by mechanical twinning. Atom positions and bonds as in Fig. 26; dotted lines for unit cells.

untwinned crystal[198,199], and for this reason the term "twin" has been enclosed in quotes[20]. It is, strictly speaking, thus perhaps more correct to refer to the twin in Fig. 62 as a type of martensite. On the other hand, when the atoms shown in Fig. 62 become disordered, a true twin is produced, and it is for this reason that the term "twin" is carried over into the ordered structure.

The "twin" shown in Fig. 62 can be visualized as being generated by the motion of twinning dislocations with a partial Burgers vector given by $b^T = \frac{1}{6}a_0[1\bar{1}1]$ on every successive ($\bar{1}12$) plane. This is the same Burgers vector that produced the stacking fault in the B2 type lattice shown in Fig. 26. It follows, therefore, that twinning and stacking faults are closely related. In this respect, it has already been shown in Fig. 26 that the disordering energy per unit area expended in moving the $\frac{1}{6}a_0[1\bar{1}1]$ twinning dislocation on the ($\bar{1}12$) plane is given by

$$\gamma_0 = 4v/\sqrt{6}\,a_0^2 \qquad (84)$$

Likewise, every partial dislocation involved in the formation of the twin shown in Fig. 62 must undergo this same expenditure of energy. Thus, for the ordered alloy, Eq. (81) must contain a fifth term given by[114]

$$E_0 = \pi \sum_i \gamma_0 R_0^2 \tag{85}$$

Since γ_0 can be equated to $\tau_0 b^T$, τ_0 can be thought of as an effective stress on the twinning dislocation due to atomic order. Combining Eqs. (83b) and (85) gives

$$-E_\tau + E_0 = -\pi \sum_i (\tau_A - \tau_0) b^T R_i^2 \tag{86}$$

Thus, atomic ordering can be thought of as giving rise to an effective stress τ_0 which reduces the effect of the applied stress τ_A.

The first experimental evidence to show that increasing the degree of atomic ordering could either impede or eliminate completely the formation of mechanical twins in an ordered alloy was obtained by Cahn and Coll with the Fe_3Al alloy[200]. Subsequent more detailed studies[38,44,161], on the other hand, showed that mechanical twinning could be induced in these alloys by either reducing the Al concentration from the stoichiometric Fe_3Al composition and/or straining to high deformation so that the flow stress was increased to that of the twinning stress, i.e. usually stage III, where the effects of slip-induced disorder could also aid the twinning process. It was also shown that mechanical "twinning" could also be induced in the Fe_3B superlattice[201–204] and that the disordering energy stored in these "twins," given by Eq. (84), could be used to "untwin" the sample, once the load was removed. In the case of Fe_3Si alloys, on the other hand, mechanical twinning was never observed in the fully ordered alloys[191], but could be detected at low temperatures in the short range ordered alloys, i.e. below Si concentrations of about 10 at. %. Reference to Table II shows that Fe_3Si, Fe_3Al, and Fe_3B are all DO_3 type superlattices whose critical ordering temperatures decrease from 1503, 848, and 620°K, respectively. It follows, therefore, that the increased propensity toward twinning with decreasing T_C is related to a decrease in the value of γ_0 in Eq. (84) due to an accompanying decrease in the value of v given by Eq. (10). These results are also in agreement with the findings for both stoichiometric[114] and non-stoichiometric[189] FeCo alloys; this shows that whereas the disordered alloys can be twinned at sufficiently low temperatures, such is not the case for the fully ordered alloys.

It is only at sufficiently low temperatures where the temperature dependent Peierls stress for slip is raised above that of the nearly temperature independent Peierls stress for twinning[205], as shown in Fig. 23, that plastic deformation will take place via twinning. Such is the case for disordered FeCo at 4.2°K where the resulting stress–strain curve is shown in

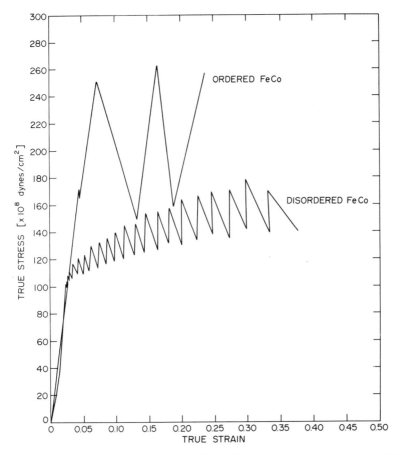

Fig. 63. Compressive stress–strain curves obtained from an ordered and a disordered FeCo alloy deformed at 4.2°K. From Fong *et al.*[*114*].

Fig. 63[*114*]. Unlike those cases where deformation occurred by slip, i.e., Figs. 13, 14, 41, 47, 52, 53, 55, and 59, Fig. 63 shows that deformation twinning occurs by a series of repeated load drops, each drop being accompanied by a sonic twin burst. Similar results have been observed for low-carbon steel tested at 4.2°K[*206*].

Deformation twinning can be favored over that of slip by increasing the strain rate[*77*]. Thus, increased strain rate may be viewed as equivalent to decreased temperature. It has, therefore, been possible to induce mechanical twinning in ordered Cu_3Au by shock loading[*133*], whereas none was found in samples deformed by conventional tensile testing. An increased propensity toward deformation twinning in ordered Fe_3Al by shock loading has also been found[207]. As mentioned earlier, it has also been shown that deformation twinning is largely responsible for the plastic deformation that occurs in the CuAu superlattice[95,108]. This may in part be due to a reported lowering of the stacking fault energy in this alloy as a result of atomic ordering[208]; however, it has also been shown that, as in the case of dislocations in these particular structures, twinning may occur in certain directions without the generation of any net disorder[20]. In concluding this section on twinning, it should also be mentioned that short- as well as long-range order will in general impede deformation twinning; however, the latter will be much more effective[189]. More detailed discussion of the relationship between mechanical twinning and atomic ordering are given in Marcinkowski[20], Marcinkowski and Fisher[77], Arunachalam and Sargent[209], and Wang[210].

Another deformation mode that appears to be intimately associated with atomic ordering is the stress-induced true martensitic transformation, i.e. those in which the positions of the atomic sites are altered as a result of the transformation. Such transformations have thus far been observed in β-brass[211], NiTi[212–217], NiAl[218,219], AuCd[220], AuZn[221], Fe_3Pt[222], Fe_3Al[223], and AgCd[221]. On the other hand, a great deal of confusion exists regarding the structure of these martensites. In addition, a "memory" effect has also been associated with these martensitic transformations; it appears to be related to the state of atomic ordering. In particular, in the case of the Fe_3Pt alloy, no memory effect could be detected for the disordered alloy[222], but did in fact appear after atomic ordering. More generally, it has been argued that plastic deformation via a martensitic transformation may be favored over that via slip and twinning due to the disordering effects associated with the latter deformation modes[223]. In the martensite transformations, the atom rearrangements involved appear to be sufficiently small so as not to generate any net disorder within the crystal[20,212,223]. As in the case of crack and twin lamellae, Eq. (81) can also be expected to be valid for martensite lamellae, except that now γ_0 in Eq. (85) should be replaced by γ_M[196]. This term reflects the volume free-energy change associated with the martensite formation and may in fact be negative. If sufficiently great, γ_M may in fact provide the necessary driving force for martensite formation, even in the absence of an applied stress τ_A.

Returning again to the subject of crack formation, it is well known that atomic ordering generally leads to increased embrittlement in alloys. This particular feature of ordered alloys represents perhaps the most important drawback in the practical application of this potential source of very high strength alloys. In particular, it has been shown that atomic ordering decreases the ductility of FeCo–2V[56,115,187,188,224–226], Fe_3Al[38,194,227], β-brass[228], and Ni_3Mn[58]. Figure 64, for example, shows the profound

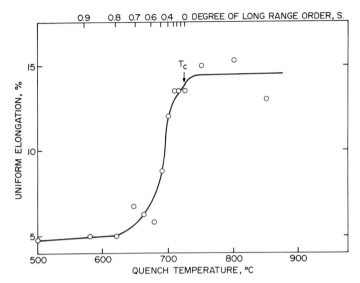

Fig. 64. The effect of atomic order on the ductility of FeCo–2V at 25°C. From Stoloff and Davies[56].

effect that atomic ordering has on the room temperature ductility of FeCo–2V[56]. In order to understand this behavior, the formulation followed in Meakin and Petch[229] will be used. Consider first the energy E_G associated with the plastic deformation of a given grain along the lines discussed in connection with Fig. 58. In particular, it is given by

$$E_G = D^2(\tau - \tau_f)^2/8\mu \tag{87}$$

The above energy may be thought of as that energy stored in the formation of a dislocation pile-up. If now a portion of this energy $\alpha^2 E_G$ is used to create a crack of length $2C$, it follows from Eq. (82) that the energy necessary to accomplish this is

$$E_{G\gamma_c} = 4\gamma_c C \tag{88}$$

Equating $\alpha^2 E_G$ with $E_{G\gamma_c}$ gives

$$\alpha^2 D^2(\tau - \tau_f)^2/8\mu = 4\gamma_c \tag{89}$$

Realizing now that the Griffith condition for crack formation is given by

$$\tau_{Gc} = (\mu\gamma_c/C)^{1/2} \tag{90}$$

it is possible to combine Eqs. (89) and (90) to obtain

$$\tau_{Gc} = 4\sqrt{2}\,\mu\gamma_c/\alpha(\tau - \tau_f)D \tag{91}$$

If now brittle fracture is defined by the condition that the yield stress $\tau_y \geq \tau_{Gc}$, then $\tau_y = \tau - \tau_f$, and employing Eq. (76), Eq. (91) becomes

$$\sigma_y k_y D^{1/2} \geq \beta \mu \gamma_c \qquad (92)$$

where $\beta = 4\sqrt{2}/\alpha$. Equation (92) is also of the same form as that obtained by Cottrell[230] and says that when the expression on the left-hand side is greater than that on the right, the yield stress σ_y will enable the crack to grow into a full fracture. It therefore follows that those factors that increase σ_y, k_y, and D relative to γ_c will promote brittle fracture. The large values of k_y for ordered FeCo compared to the disordered alloy could thus account in part for the increased brittleness. The question of γ_c will be taken up presently.

The effect of atomic order on ductility can also be seen by reference to Fig.65 which shows the percent elongation versus composition for a series of fully ordered FeAl alloys[38,194]. These alloys correspond to those from

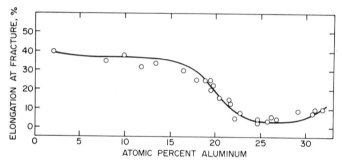

Fig. 65. Variation of ductility with Al concentration in fully ordered FeAl alloys (477°K). From Leamy[38].

which the curves of Fig. 60 were obtained. It is immediately apparent from this figure that the marked decrease in ductility corresponds to the higher amounts of long-range order associated with the stoichiometric Fe_3Al composition. A large part of this loss in ductility is again anticipated to be due to the increase in k_y with atomic order in Eq. (92).

A study of the ductile–brittle transition has also been carried out for both the fully ordered as well as the disordered FeCo-2V alloys[115]. These results are shown in Fig. 66. The nearly 500°C increase in the ductile–brittle transition temperature resulting from atomic ordering is quite dramatic. The low transition temperature associated with the disordered alloy may be associated in part with the relatively low value of k_y coupled with the relatively strong temperature dependence of σ_y shown in Fig. 30. On the other hand, the high ductile–brittle transition temperature characteristic of the ordered alloy may in part be due to the high value of k_y coupled with its strong temperature dependence[187,188].

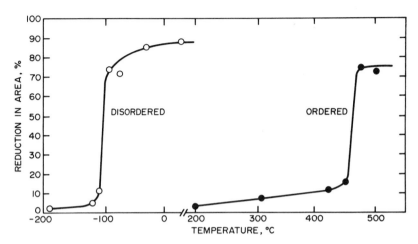

Fig. 66. Effect of atomic order on the ductile–brittle transition of FeCo–2V. From Johnston *et al.*[*115*].

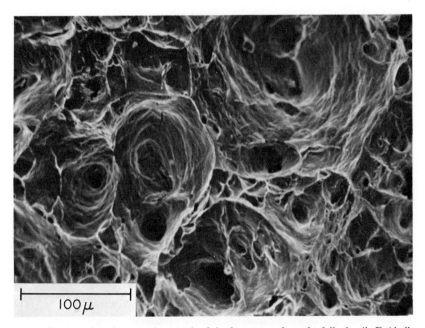

Fig. 67a. Scanning electron micrograph of the fracture surface of a fully ductile FeAl alloy containing 2 at. % Al, illustrating fracture by void coalescence. From Marcinkowski *et al.*[*227*].

Attention will now be focused on the relationship between γ_c and atomic ordering. It has been shown that atomic ordering leads to intercrystalline crack formation in FeCo[114,231,232]. On the other hand, in the case of ordered FeCo-2V, it has been shown that the fracture is essentially of the transgranular cleavage type[224–226] at low temperatures and reverts to intercrystalline at higher temperatures. Similar types of observations have been made for β-brass[233,234]. On the other hand, fracture in NiAl[138,235], FeAl[231], and Ni$_3$Al[235] is entirely intercrystalline. The detailed manner in which atomic ordering affects the fracture mode can be seen by reference to the scanning electron micrograph shown in Fig. 67[227]. These fracture surfaces corresponded to the FeAl alloys shown in Figs. 60 and 65. In particular, the fracture surface shown in Fig. 67a consisted en- tirely of dimples brought about by void coalescence and were characteristic of the ductile fractures characteristic of the alloys containing from 0 to 8 at. % Al. From 8 to 21 at. % Al, the dimpled surface gives way to trans- granular cleavage as shown in Fig. 67b, while from 21 to 26 at. % Al, the transgranular cleavage gives way to intergranular cleavage as shown in

Fig. 67b. Scanning electron micrograph of the fracture surface in an embrittled FeAl alloy containing 21.8 at. % Al, showing predominant transgranular fracture. From Marcinkowski et al.[227].

Fig. 67c. Scanning electron micrograph of the fracture surface in a fully embrittled iron–aluminum alloy containing 25.5 at. % Al, exhibiting intergranular fracture. From Marcinkowski *et al.*[227].

Fig.67c[38,161,227]. It has been argued that intergranular fracture in many ordered alloys such as AgMg, NiAl, and NiGa is due to the segregation of impurity atoms to the grain boundary[236,237]. An alternative proposal for this intergranular embrittlement, based upon the inherent nature of a grain boundary itself within an ordered alloy, has also been proposed. In particular, Fig. 68 shows an 18.9° symmetric tilt boundary in a simple ordered structure[238]. The boundary is seen to consist of a vertically aligned array of edge dislocations. Incorrect atom neighbors are observed to exist across the boundary between every second pair of dislocations. In a simple sense, therefore, the grain boundary may be considered to consist of a vertically aligned array of superlattice dislocations. This same reasoning obtains for low angle tilt[239] and twist[240] boundaries as well as for high angle tilt[238] and twist[241] boundaries. In particular, Fig. 69 shows a transmission electron micrograph of a vertically aligned array of superlattice dislocations comprising a 47' low angle symmetric tilt boundary in a fully ordered FeCo alloy[242,243]. It is a simple matter to show that the vertical extension

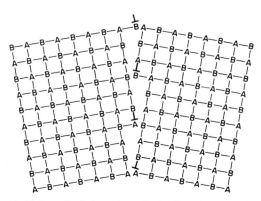

Fig. 68. Symmetric high angle tilt boundary of misorientation angle $\theta = 18.9°$ in a simple ordered lattice. From Marcinkowski *et al.*[238].

of the superlattice dislocation r_y within a symmetric tilt boundary of infinite extent is given by[239]

$$r_y = (h/\pi) \cot^{-1} \left[2h(1 - v)\gamma_0/\mu b^2 \right] \tag{93}$$

where h is the spacing between superlattice dislocations. Note also that as the tilt angle θ increases, i.e. as h increases r_y/h in Eq. (93) approaches 0.5, as shown in Fig. 68. On the other hand, for very small θ angles, i.e. large h, the bracketed term in Eq. (93) can be written as $\mu b^2/[2h(1 - v)\gamma_0]$ so that this equation then reduces to

$$r_y = \frac{\mu b^2}{2\pi\gamma_0(1 - v)} \tag{94}$$

which is identical to the expression given by Eq. (7) for a single superlattice dislocation. The basis for the vertical alignment of the superlattice dislocations within a symmetric tilt boundary, such as shown in Figs. 68 and 69 arises from the $\sin^2 \theta$ term in Eq. (4b). In particular, for $\theta = \pi/2$, i.e. vertical alignment, the interaction energy between a pair of ordinary edge dislocations of the same sign is a minimum.

The above analysis appears at first sight to be somewhat at odds with the earlier quoted observations of transgranular fracture in some ordered alloys. However, this transgranular fracture may be a manifestation of the presence of a large number of low angle boundaries in these particular alloys. These low angle boundaries are invariably attached to the higher angle boundaries as can be seen in the transmission electron micrograph of Fig. 70 obtained from a fully ordered FeCo alloy[243]. The fracture path will thus progress through both the high and low angle boundaries, and it is the latter paths that give rise to the transgranular nature of the fracture surfaces. Because of the disorder associated with low angle boundaries in superlattices, they are

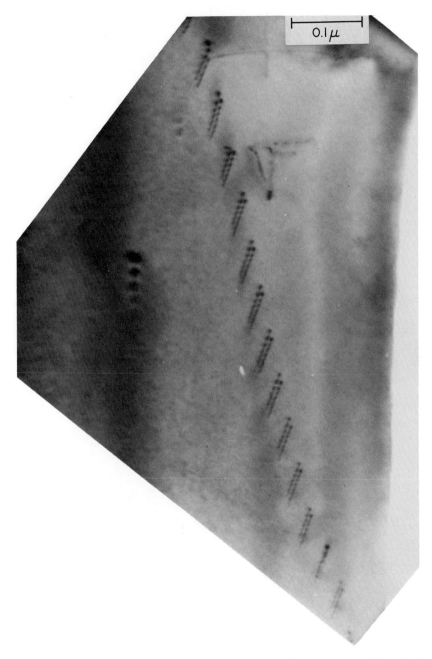

Fig. 69. Transmission electron micrograph of a symmetric tilt boundary in a fully ordered FeCo alloy. From Tseng[243].

2 μ

Fig. 70. Dark field transmission electron micrograph composite obtained from a fully ordered FeCo alloy showing low angle boundaries terminating on high angle boundary. From Tseng[*243*].

of relatively high energy, particularly when the critical ordering temperature T_C is large, and thus they can be minimized or eliminated by a suitable heat treatment. Thus, it would appear that higher degrees of order, as in the case of the Fe_3Al alloys in Fig. 67, or higher values of T_C, as in the case of FeAl and NiAl, would minimize the number of low angle boundaries and in turn favor intergranular fracture, in agreement with the previous observations. Also pertinent to the above arguments is the observation that plastic deformation, followed by subsequent partial recrystallization, induces marked ductility in ordered FeCo–2V alloys[225,244]. This is in keeping with the prediction that such processes induce a very high density of low angle boundaries within these alloys, which deviates the fracture path from the high angle to the low angle boundaries. Frequent deviations of this type are expected to give rise to a marked increase in the crack surface energy γ_c. Thus, it may be argued in general that atomic ordering leads to a reduction of γ_c in Eq. (92), which in turn promotes brittle fracture.

There appear to be at least two exceptions to the general argument that atomic ordering increases the brittleness of alloys. The first occurs in the case of Mg_3Cd where atomic ordering is found to increase the ductility[81]. It has been argued that this behavior arises from a reduction of k_y in Eq. (92) which results from a decrease in the mean orientation factor $\bar{m}$ through the relation $k = \bar{m}^2 r^{1/2} \tau_c$. The decrease in $\bar{m}$ is brought about by an increase in the number of slip systems occasioned by atomic order[224].

In the case of the Cu_3Au superlattice, atomic ordering appears to have essentially no effect on the ductility of the disordered alloy, and both states possess high ductility. Again, this may be due to the low value of k_y which results from the ease of cross slip in this alloy. In addition, however, T_C, as can be seen from Table II, is rather low, so that the effects of order, particularly upon γ_c in Eq. (92), could be somewhat small.

In concluding this section on fracture, it should be noted that the fracture stress of FeCo–2V was not observed to vary as $C^{-1/2}$, as might be expected from Eq. (90)[187,188]. Since Eq. (92) is based upon Eq. (90), it has been argued that the theoretical predictions garnered from Eq. (92) may be incorrect. It follows, therefore, that further work in this area will be required before this question is satisfactorily resolved. It is also important to note that if the cracks are hindered in their propagation, as might occur during a compression test, the stress–strain curve appears qualitatively similar to that obtained for twinning. This can be seen by reference to the compressive stress–strain curve obtained for the fully ordered FeCo alloy tested at 4.2°K[114]. Atomic order raised the stress for twinning, in accordance with Eq. (86), so as to make crack formation the easier mode of deformation. The load drops associated with the ordered alloy in Fig. 63 thus correspond to bursts of crack lamellae similar to the bursts of twin lamellae that occurred in the corresponding disordered alloy.

V. Effects of Atomic Order on Fatigue and Creep Behavior

It has been demonstrated that atomic ordering significantly increases the fatigue properties of FeCo–2V and Ni_3Mn[245]. This can be seen by reference to Fig. 71 which shows the fatigue curves for ordered and disordered Ni_3Mn[245]. These data show that atomic ordering gives rise to an increase in the fatigue strength of the alloy. Similar results have been observed for FeSi alloys as the percentage of Si is increased[246]. In all of these alloys, the improvement of fatigue strength has been accompanied by a decreased propensity toward cross slip and a parallel decrease in slip band generation. This is in agreement with the postulate that ease of cross slip is a prerequisite for the generation of a slip band[247]. Slip band generation, on the other hand, is an important prerequisite for the formation of surface extrusions and

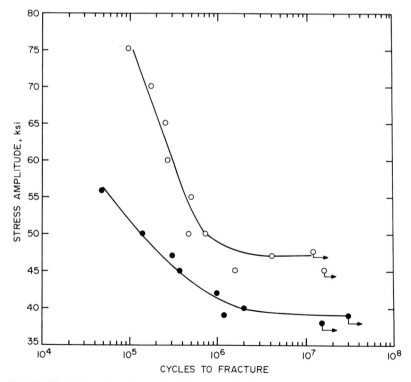

Fig. 71. Effect of atomic order on the fatigue strength of Ni_3Mn; 74.5 w/o Ni–25.5 w/o Mn; ○ ordered; ● quenched; 25°C; yield stress (ksi)/tensile stress (ksi): Ordered, 42/114; quenched, 27/102. From Boettner et al.[245].

intrusions, which may act as the nuclei for crack nucleation. The above findings are also supported by the observations that atomic ordering gives rise to a very uniform distribution of fine slip lines, in contrast to the slip bands characteristic of the corresponding disordered alloys[59,60,248–250]. In addition, for those alloys thus far examined, i.e. FeCo[245], Ni₃Mn[245], and β-brass[251], it has been shown that the fatigue cracks are intergranular in nature, in agreement with the arguments presented in the previous section.

The final topic to be considered in the present review concerns the effect of atomic order on the creep behavior of alloys. Herman and Brown[252] were the first to show that atomic ordering in β-brass leads to a marked increase in the steady-state creep rate of this alloy. These results can be seen in Fig. 72

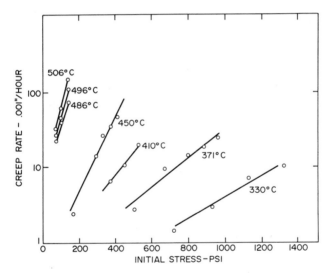

Fig. 72. Effect of atomic order on the steady-state creep behavior of β-brass. From Herman and Brown[252].

where the creep rate is seen to undergo a sharp decrease below the critical ordering temperature of $T_c = 465°C$. Similar results have also been found in subsequent studies of β-brass[253,254] as well as for Ni₃Fe[255], Ni₃Al[148], NiAl[256,257], Fe₃Al alloys[258], MgCd[259], and FeSi alloys[260].

The first successful theory of creep in ordered alloys was formulated by Flinn[148]. In this theory it was assumed that whereas at low temperatures, superlattice dislocations would be extended within their slip plane, this might not be the lowest energy configuration. In particular, it was argued that the superlattice dislocation might lower its energy by extending normal to the slip plane, which it could accomplish by climb at elevated temperatures. The original justification for vertical superlattice extension was based upon the calculations by Flinn[148] which predicted that the APB energy in planes inclined to the slip plane was of lower energy than APBs within the slip plane. Although this is apparently the case for the $L1_2$ type superlattices, as can be seen in connection with the discussion made with respect to Fig. 40, it has been argued by Brown and Lenton[253] that Flinn's calculations were in error for B2 type superlattices, and that the APB energy is in fact a minimum within the {110} glide plane. These results have in turn also been argued to be incorrect, and that much more refined calculations of APB energies are necessary before the above problem can be resolved[261,262]. Aside from APB considerations, however, it has already been reasoned that on the basis of elastic interaction energies alone between a pair of edge dislocations of the same sign, vertical superlattice extension would be favored. This argument by itself is sufficient to restore the basis of Flinn's theory, to say nothing of elastic anisotropy effects which could also be important.

In order for a vertically extended superlattice dislocation to glide without the generation of APBs, such as shown schematically in Fig. 73a, it is necessary that the APBs follow the gliding dislocations, as indicated in Fig. 73b.

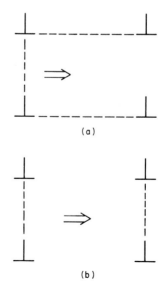

(a)

(b)

Fig. 73. Glide of vertically extended superlattice dislocation (a) at low temperatures, with the generation of antiphase boundaries, (b) at high temperatures, without the generation of antiphase boundaries.

Since the applied stress τ is less than that required to produce APBs, the dislocations shown in Fig. 73b will move forward only if atoms in front of the dislocation jump so as to produce wrong bonds. The jump rate for forward motion will then be given by

$$J_1 = v' \exp\left(-\frac{\Delta F_1}{RT}\right) \exp\left[-\frac{(\Delta F_2 + \varepsilon)}{RT}\right] \exp\left(\frac{\tau b}{kT}\right) \tag{95a}$$

while the jump rate for backward motion is

$$J_2 = v' \exp\left(-\frac{\Delta F_1}{RT}\right) \exp\left[-\frac{(\Delta F_2 + \varepsilon)}{RT}\right] \exp\left(-\frac{\tau b}{kT}\right) \tag{95b}$$

where ΔF_1 and ΔF_2 are the free energy changes associated with vacancy creation and motion, respectively, while ε is the energy needed to form wrong bonds, and v' is the vibrational frequency. Assuming that $\tau b \ll kT$, the net jump rate J can be written as

$$J = J_1 - J_2 = \frac{2\tau b}{kT} v' \exp\left(\frac{\Delta S}{R}\right) \exp\left(-\frac{\Delta H + \varepsilon}{RT}\right) \tag{96}$$

where ΔS and ΔH are the activation entropy and heat of diffusion, respectively. Since[263]

$$D_0 = a^2 v' \exp(\Delta S/R) \tag{97}$$

Eq. (96) becomes

$$J = \frac{\sqrt{2}\tau}{akT} D_0 \exp\left(-\frac{Q}{RT}\right) \tag{98}$$

where $Q = \Delta H + \varepsilon$ and $b = \sqrt{2} a$ for a face-centered cubic lattice. The jump rate expressed by Eq. (98) will give rise to a dislocation velocity of

$$\frac{dx}{dt} = Jb = \frac{2\tau}{kT} D_0 \exp\left(-\frac{Q}{RT}\right) \tag{99}$$

When substituted into Weertman's equation for microcreep, the following expression for creep rate obtains:

$$\dot{\beta} = \left[\frac{2\pi D_0 b}{\mu^2 kT}\right] \sigma^3 \exp\left(-\frac{Q}{RT}\right) \tag{100}$$

The creep rate data for Ni_3Al shown in Fig. 74 are seen to be in good agreement with Eq. (100).

In concluding the present review, it is to be emphasized that no attempt has been made to treat the mechanical properties of two-phase systems, one or both phases of which are atomically ordered. The processes involved in

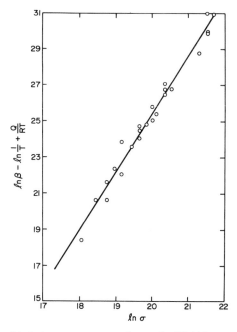

Fig. 74. Relationship between creep rate and stress for Ni_3Al in accordance with Eq. (99) in text. From Flinn[148].

these alloys are expected to be much more complex than those characteristic of the single phase alloys discussed here. These multiphase alloys, however, are extremely important from a technical point of view and form the basis of the high-temperature superalloys. Extensive reviews on this subject have recently been prepared and are listed in Kear *et al.*[10], Sims and Hagel[145], and Oblak and Kear[264].

VI. Summary and Conclusions

An extensive review has been carried out with respect to the effects of atomic ordering on the mechanical properties of alloys. Emphasis has been placed on the micromechanisms, i.e. dislocation mechanisms, associated with mechanical behavior. It has been shown that the most distinguishing feature responsible for the marked changes in mechanical properties that accompany atomic ordering is the slip-produced antiphase boundary. It is this antiphase boundary energy that is responsible for the modification of the dislocation morphology. The net result is that a marked increase in work-hardening rate accompanies atomic ordering. In addition, the grain boundary effects are markedly accentuated, while the creep and fatigue strengths are greatly improved. On the other hand, the ductility is markedly decreased as a result of atomic order. Aside from a study of the effects of atomic order in its own right, it has been demonstrated that atomic ordering provides a very powerful tool with which to verify the presently existing theories relating to mechanical behavior as well as to formulate new ones.

ACKNOWLEDGMENTS

The author would like to express his sincere appreciation to Dr. K. Sadananda of The Engineering Materials Group and The Department of Mechanical Engineering of The University of Maryland for numerous discussions dealing with all aspects of atomic ordering. Acknowledgment is also made to the United States Atomic Energy Commission for its support of the present study which was carried out under Contract No. AT-(40-1)-3935.

References

1. G. Sachs and J. Weerts, *Z. Phys.* **67**, 507 (1931).
2. E. C. Bain, *Chem. Met. Eng.* **28**, 65 (1923).
3. G. Tammann, *Z. Anorg. Allg. Chem.* **107**, 1 (1919).
4. F. C. Nix and W. Shockley, *Rev. Mod. Phys.* **10**, 1 (1938).
5. H. Lipson, *Prog. Metal Phys.* **2**, 1 (1950).
6. T. Muto and Y. Takagi, *Solid State Phys.* **1**, 193 (1955).
7. J. H. Westbrook, ed., "Intermetallic Compounds." Wiley, New York, 1967.
8. N. S. Stoloff and R. G. Davies, *Prog. Mater. Sci.* **13**, 1 (1966).
9. J. H. Westbrook, ed., "Mechanical Properties of Intermetallic Compounds." Wiley, New York, 1960.
10. B. H. Kear, C. T. Sims, N. S. Stoloff, and J. H. Westbrook, eds., "Ordered Alloys—Structural Application and Physical Metallurgy." Claitor's Publ. Div., Baton Rouge, Louisiana, 1970.
11. M. J. Marcinkowski, *in* "Electron Microscopy and Strength of Crystals" (G. Thomas and J. Washburn, eds.), p. 333. Wiley (Interscience), New York, 1963.
12. J. S. Koehler and F. Seitz, *J. Appl. Mech.* **14**, A217 (1947).
13. M. J. Marcinkowski and R. M. Fisher, *Electron Microsc., Proc. Sec. Eur. Reg. Conf., Delft* **1**, 400 (1960).
14. M. J. Marcinkowski, R. M. Fisher, and N. Brown, *J. Appl. Phy.* **31**, 1303 (1960).
15. M. J. Marcinkowski, N. Brown, and R. M. Fisher, *Acta Met.* **9**, 129 (1961).
16. N. Brown and M. Herman, *Trans. AIME* **206**, 1353 (1956).
17. J. P. Hirth and J. Lothe, "Theory of Dislocations." McGraw-Hill, New York, 1968.
18. M. J. Marcinkowski, *Advan. Mater. Res.* **5**, 443 (1971).
19. L. Guttman, *Solid State Phys.* **3**, 145 (1956).
20. M. J. Marcinkowski, *Mem. Sci. Rev. Met.* **69**, 579 (1972).
21. W. A. Rachinger and A. H. Cottrell, *Acta Met.* **4**, 109 (1956).
22. A. H. Cottrell, "Relation of Properties of Microstructure," p. 131. Amer. Soc. Metals, Cleveland, Ohio.
23. A. Ball and R. E. Smallman, *Acta Met.* **14**, 1517 (1966).
24. E. P. Lautenschlager, T. Hughes, and J. O. Brittain, *Acta Met.* **15**, 1347 (1967).
25. D. I. Potter, *J. Mater. Sci. Eng.* **5**, 201 (1970).
26. H. J. Leamy, F. X. Kayser, and M. J. Marcinkowski, *Phil. Mag.* **20**, 779 (1969).

27. R. T. Pascoe and C. W. A. Newey, *Phys. Status Solidi* **29**, 357 (1968).
28. E. M. Schulson and E. Teghtsoonian, *Phil. Mag.* **19**, 155 (1969).
29. A. K. Head, *Phys. Status Solidi* **19**, 185 (1967).
30. C. Zener, "Elasticity and Anelasticity of Metals." Univ. of Chicago Press, Chicago, Illinois, 1948.
31. A. K. Head, M. H. Loretto, and P. Humble, *Phys. Status Solidi* **20**, 521 (1967).
32. C. H. Lloyd and M. H. Loretto, *Phys. Status Solidi* **39**, 163 (1970).
33. M. H. Loretto and R. J. Wasilewski, *Phil. Mag.* **23**, 1311 (1971).
34. G. E. Lakso and M. J. Marcinkowski, *Trans. AIME* **245**, 1111 (1969).
35. J. D. Eshelby, *Phil. Mag.* **40**, 903 (1949).
36. A. J. E. Foreman, *Acta Met.* **3**, 322 (1955).
37. A. H. Cottrell, "Dislocations and Plastic Flow in Crystals." Oxford Univ. Press (Clarendon), London and New York, 1953.
38. H. J. Leamy, Ph.D. Thesis, Iowa State Univ., Ames, 1967.
39. M. J. Marcinkowski and H. J. Leamy, *Phys. Status Solidi* **24**, 149 (1967).
40. E. Kröner, "Kontinuumstheorie der Versetzungen und Eigenspannungen." Springer, Berlin, 1958.
41. M. J. Marcinkowski and K. S. Sree Harsha, *J. Appl. Phys.* **39**, 1775 (1968).
42. M. J. Marcinkowski, *Advan. Mater. Res.* **5**, 443 (1971).
43. M. F. Ashby, *Acta Met.* **14**, 680 (1966).
44. H. J. Leamy, F. X. Kayser, and M. J. Marcinkowski, *Phil. Mag.* **20**, 763 (1969).
45. L. K. France, C. S. Hartley, and C. N. Reid, *Metal Sci. J.* **1**, 65 (1967).
46. P. Chaudhari, *Acta Met.* **14**, 69 (1966).
47. M. J. Marcinkowski and G. E. Lakso, *J. Appl. Phys.* **38**, 2124 (1967).
48. K. Sadananda and M. J. Marcinkowski, *J. Appl. Phys.* **43**, 2609 (1972).
49. K. Sadananda and M. J. Marcinkowski, *J. Appl. Phys.* **43**, 293 (1972).
50. K. Sadananda and M. J. Marcinkowski, *J. Mater. Sci.* **8**, 839 (1973).
51. K. Sadananda and M. J. Marcinkowski, *Mater. Sci. Eng.* **11**, 51 (1973).
52. K. Sadananda and M. J. Marcinkowski, *Cryst. Lattice Defects* **3**, 177 (1972).
53. M. J. Marcinkowski and H. J. Leamy, *J. Appl. Phys.* **40**, 3095 (1969).
54. S.-T. Fong, K. Sadananda, and M. J. Marcinkowski, *Met. Trans.* **5**, 1239 (1974).
55. M. J. Marcinkowski and H. Chessin, *Phil. Mag.* **10**, 837 (1964).
56. N. S. Stoloff and R. G. Davies, *Acta Met.* **12**, 473 (1964).
57. P. Moine, J. P. Eymery, and P. Grosbras, *Phys. Status Solidi (B)* **46**, 177 (1971).
58. M. J. Marcinkowski and D. S. Miller, *Phil. Mag.* **6**, 871 (1961).
59. F. M. C. Besag and R. E. Smallman, in "Ordered Alloys—Structural Application and Physical Metallurgy" (B. H. Kear, C. T. Sims, N. S. Stoloff, and J. H. Westbrook, eds.), p. 259. Claitor's Publ. Div., Baton Rouge, Louisiana, 1970.
60. F. M. C. Besag and R. E. Smallman, *Acta Met.* **18**, 429 (1970).
61. A. Lawley, E. A. Vidoz, and R. W. Cahn, *Acta Met.* **9**, 287 (1961).
62. H. Green and N. Brown, *Trans. AIME* **197**, 1240 (1953).
63. N. Brown, in "Mechanical Properties of Intermetallic Compounds" (J. H. Westbrook, ed.), p. 177. Wiley, New York, 1960.
64. N. Brown, *Phil. Mag.* **4**, 693 (1959).
65. J. B. Cohen and M. E. Fine, *J. Phys. Radium* **23**, 749 (1962).
66. J. B. Cohen and M. E. Fine, *Acta Met.* **11**, 1106 (1963).
67. M. J. Marcinkowski and L. Zwell, *Acta Met.* **11**, 373, (1963).
68. M. J. Blackburn, *Trans. AIME* **239**, 660 (1967).
69. W. Bell, W. R. Roser, and G. Thomas, *Acta Met.* **12**, 1247 (1964).
70. H. E. Cook, *Trans. AIME* **242**, 1599 (1968).
71. P. Moine and P. Grosbras, *C. R. Acad. Sci., Ser. B* **268**, 759 (1969).

72. P. S. Rudman, *Acta Met.* **10**, 253 (1962).
73. P. A. Flinn, "Strengthening Mechanisms in Solids," p. 17. Amer. Soc. Metals, Metals Park, Ohio.
74. J. C. Fisher, *Acta Met.* **2**, 9 (1954).
75. N. Brown, *Acta Met.* **7**, 210 (1959).
76. K. Sumino, *Sci. Rep. Res. Inst. Tohoku Univ., Ser. A* **10**, 283 (1958).
77. M. J. Marcinkowski and R. M. Fisher, *J. Appl. Phys.* **34**, 2135 (1963).
78. M. J. Marcinkowski and N. Brown, *Acta Met.* **9**, 764 (1961).
79. J. B. Mitchell, O. Abo-El-Fotoh, and J. E. Dorn, *Met. Trans.* **2**, 3265 (1971).
80. M. O. Aboelfotoh, *Phys. Status Solidi (A)* **14**, 545 (1972).
81. R. G. Davies and N. S. Stoloff, *Trans. AIME* **230**, 390 (1964).
82. N. S. Stoloff and R. G. Davies, *ASM (Amer. Soc. Metals) Trans. Quart.* **57**, 247 (1964).
83. A. Seeger, *Phil. Mag.* **46**, 1194 (1955).
84. G. C. Kuczynski, R. F. Hochman, and M. Doyama, *J. Appl. Phys.* **26**, 871 (1955).
85. M. Hirabayashi and S. Weissmann, *Acta Met.* **10**, 25 (1962).
86. B. Hansson and R. S. Barnes, *Acta Met.* **12**, 315 (1964).
87. J. B. Newkirk, A. H. Geisler, D. L. Martin, and R. Smoluchowski, *J. Metals* **188**, 1249 (1950).
88. J. B. Newkirk, R. Smoluchowski, A. H. Geisler, and D. L. Martin, *J. Appl. Phys.* **22**, 290 (1951).
89. J. B. Newkirk, R. Smoluchowski, A. H. Geisler, and D. L. Martin, *Acta Crystallogr.* **4**, 507 (1951).
90. H. Lipson, D. Shoenberg, and G. V. Stuport, *J. Inst. Metals* **67**, 333 (1941).
91. L. E. Tanner and M. F. Ashby, *Phys. Status Solidi (B)* **33**, 59 (1969).
92. B. A. Greenberg, *Phys. Status Solidi (B)* **42**, 459 (1970).
93. V. I. Syutkina and E. S. Yakovleva, *Phys. Status Solidi* **21**, 465 (1967).
94. B. A. Grinberg, V. I. Syutkina, and E. S. Yakovleva, *Phys. Metals Metallogr. (USSR)* **26**(1), 17 (1968) [*Fiz. Metal. Metalloved.* **26**(1), 18 (1968)].
95. D. W. Pashley, J. L. Robertson, and M. J. Stowell, *Phil. Mag.* **19**, 83 (1969).
96. R. C. Crawford, I. L. F. Ray, and D. J. H. Cockayne, *Phil. Mag.* **27**, 1 (1973).
97. I. L. F. Ray, R. C. Crawford, and D. J. H. Cockayne, *Phil. Mag.* **21**, 1027 (1970).
98. R. J. Taunt, Ph.D. Thesis, Univ. of Cambridge, Cambridge, England, 1973.
99. G. W. Ardley, *Acta Met.* **3**, 525 (1955).
100. R. G. Davies, *Trans. AIME* **230**, 903 (1964).
101. Y. P. Selisskii, *Phys. Metals Metallogr. (USSR)* **11**(1), 124 (1961) [*Fiz. Metal. Metalloved.* **11**(1), 128 (1961)].
102. R. G. Davies and N. S. Stoloff, *Acta Met.* **11**, 1347 (1963).
103. N. A. Koneva and A. D. Korotayev, *Phys. Metals Metallogr. (USSR)* **21**(1), 53 (1966) [*Fiz. Metal. Metalloved.* **21**(1), 54 (1966)].
104. M. J. Hordon, *Trans. AIME* **227**, 260 (1963).
105. W. D. Biggs and T. Broom, *Phil. Mag.* **45**, 246 (1954).
106. L. Nowack, *Z. Metallk.* **22**, 94 (1930).
107. E. W. Horne and E. A. Starke, Jr., *Phil. Mag.* **23**, 741 (1971).
108. V. S. Arunachalam and R. W. Cahn, *J. Mater. Sci.* **2**, 160 (1967).
109. B. Ramaswami, *Scr. Met.* **4**, 865 (1970).
110. V. S. Arunachalam, *Scr. Met.* **4**, 859 (1970).
111. T. Chandra and B. Ramaswami, *Scr. Met.* **4**, 175 (1970).
112. H. J. Logie, *Acta Met.* **5**, 106 (1957).
113. D. W. Pashley and A. E. B. Presland, *J. Inst. Metals* **87**, 419 (1958–1959).
114. S.-T. Fong, M. J. Marcinkowski, and K. Sadananda, *Acta Met.* **21**, 799 (1973).
115. T. L. Johnston, R. G. Davies, and N. S. Stoloff, *Phil. Mag.* **12**, 305 (1965).

116. T. G. Langdon and J. E. Dorn. *Phil. Mag.* **17**, 999 (1968).
117. D. P. Pope, *Phil. Mag.* **25**, 917 (1972).
118. D. P. Pope, *Phil. Mag.* **27**, 541 (1973).
119. R. G. Davies and N. S. Stoloff, *Phil. Mag.* **12**, 297 (1965).
120. J. H. Kirby and F. W. Noble. *Phil. Mag.* **19**, 877 (1969).
121. F. W. Noble and J. H. Kirby, *in* "Ordered Alloys–Structural Applications and Physical Metallurgy" (B. H. Kear, C. T. Sims, N. S. Stoloff, and J. H. Westbrook, eds.), p. 321. Claitor's Publ. Div., Baton Rouge, Louisiana, 1970.
122. P. Guyot and J. E. Dorn, *Can. J. Phys.* **45**, 983 (1967).
123. R. A. Foxall, M. S. Duesbery, and P. B. Hirsch, *Can. J. Phys.* **45**, 607 (1967).
124. F. Kroupa and V. Vitek, *Can. J. Phys.* **45**, 945 (1967).
125. M. J. Marcinkowski, to be published (1974).
126. L. E. Popov, E. V. Kozlov, and I. V. Tereshko, *Phys. Metals Metallogr. (USSR)* **26**(4), 129 (1968) [*Fiz. Metal. Metalloved.* **26**(4), 709 (1968)].
127. O. Dahl, *Z. Metallk.* **28**, 133 (1936).
128. R. Hahn and E. Kneller, *Z. Metallk.* **49**, 480 (1958).
129. F. E. Jaumot, Jr. and A. Swatzky, *Acta Met.* **4**, 127 (1956).
130. J. B. Cohen and M. B. Bever, *Trans. AIME* **218** 155 (1960).
131. P. Beardmore, A. H. Holtzmann, and M. B. Bever, *Trans. AIME* **230**, 725 (1964).
132. J. J. West, S. G. Cupschalk, and F. A. Dahlman, Jr., *Trans. AIME* **236**, 421 (1966).
133. D. E. Mikkola and J. B. Cohen, *Acta Met.* **14**, 105 (1966).
134. G. K. Wertheim and J. H. Wernick, *Acta Met.* **15**, 297 (1967).
135. L. E. Popov, E. V. Kozlov, and N. A. Aleksandrov, *Phys. Status Solidi* **13**, K105 (1966).
136. L. E. Popov, E. V. Kozlov, N. A. Aleksandrov, and S. G. Shteyn, *Phys. Metals Metallogr. (USSR)* **21**(6), 107 (1966) [*Fiz. Metal. Metalloved.* **21**(6), 920 (1966)].
137. R. J. Wasilewski, S. R. Butler, and J. E. Hanlon, *Trans. AIME* **239**, 1357 (1967).
138. A. Ball and R. E. Smallman, *Acta Met.* **14**, 1349 (1966).
139. R. T. Pascoe and C. W. Newey, *Metal Sci. J.* **2**, 138 (1969).
140. A. G. Rozner and R. J. Wasilewski, *J. Inst. Metals* **94**, 169 (1966).
141. A. R. Causey and E. Teghtsoonian, *Met. Trans.* **1**, 1177 (1970).
142. J. C. Terry and R. E. Smallman, *Phil. Mag.* **8**, 1827 (1963).
143. A. K. Mukherjee and J. E. Dorn, *Trans. AIME* **230**, 1065 (1964).
144. V. B. Kurfman, *Acta Met.* **13**, 307 (1965).
145. C. T. Sims and W. C. Hagel, eds., "The Superalloys." Wiley, New York, 1972.
146. R. G. Davies and N. S. Stoloff, *Trans. AIME* **233**, 714 (1965).
147. P. H. Thornton, R. G. Davies, and T. L. Johnston, *Met. Trans.* **1**, 207 (1970).
148. P. A. Flinn, *Trans. AIME* **218**, 145 (1960).
149. S. M. Copley and B. H. Kear, *Trans. AIME* **239**, 977 (1967).
150. P. H. Thornton and R. G. Davies, *Met. Trans.* **1**, 549 (1970).
151. S. Takeuchi and E. Kuramoto, *Acta Met.* **21**, 415 (1973).
152. S. Takeuchi and E. Kuramoto, *J. Phys. Soc. Jap.* **31**, 1282 (1971).
153. B. H. Kear and H. G. F. Wilsdorf, *Trans. AIME* **224**, 382 (1962).
154. B. H. Kear, *Acta Met.* **12**, 555 (1964).
155. B. H. Kear, *Acta Met.* **14**, 659 (1966).
156. M. J. Marcinkowski and D. E. Campbell, *in* "Ordered Alloys—Structural Applications and Physical Metallurgy" (B. H. Kear, C. T. Sims, N. S. Stoloff, and J. H. Westbrook, eds.), p. 331. Claitor's Publ. Div., Baton Rouge, Louisiana, 1970.
157. R. W. Guard and J. H. Westbrook, *Trans. AIME* **215**, 807 (1959).
158. H. Conrad, *in* "High-Strength Materials" (V. F. Zackay, ed.), p. 436. Wiley, New York, 1965.

159. G. Lakso and M. J. Marcinkowski, *Electron Microsc., Proc. Int. Congr., 6th Kyoto,* Vol. I, p. 309 (1966).
160. R. G. Davies and N. S. Stoloff, *Acta Met.* **11,** 1187 (1963).
161. H. J. Leamy and F. X. Kayser, *Phys. Status Solidi* **34,** 765 (1969).
162. R. G. Davies and N. S. Stoloff, *Phil. Mag.* **9,** 349 (1964).
163. J. Czernichow and M. J. Marcinkowski, *Met. Trans.* **2,** 3217 (1971).
164. M. A. Audew, M. P. Victoria, and A. E. Vidoz, *Phys. Status Solidi* **31,** 697 (1969).
165. M. Victoria and A. E. Vidoz, *Acta Met.* **15,** 676 (1967).
166. M. P. Victoria and A. E. Vidoz, *Phys. Status Solidi* **28,** 131 (1968).
167. A. C. Arko and Y. H. Liu, *Met. Trans.* **2,** 1875 (1971).
168. L. A. Gerzha, V. I. Syutkina, and E. S. Yakovleva, *Phys. Metals Metallogr. (USSR)* **20**(3), 115 (1965) [*Fiz. Metal. Metalloved.* **20**(3), 433 (1965)].
169. J. H. Kirby and F. W. Noble, *Phil. Mag.* **16,** 1009 (1967).
170. L. A. Gerzha, V. A. Syutkina, and E. S. Yakovleva, *Phys. Metals Metallogr. (USSR)* **18**(5), 125 (1964) [*Fiz. Metal. Metalloved.* **18**(5), 770 (1964)].
171. A. E. Vidoz and L. N. Brown, *Phil. Mag.* **7,** 1167 (1962).
172. G. Schoeck, *Acta Met.* **17,** 147 (1969).
173. A. E. Vidoz, *Phys. Status Solidi* **28,** 145 (1968).
174. G. Schoeck and E. Perez, *Scr. Met.* **5,** 421 (1971).
175. L. I. Vasil'Yev and A. N. Orlov, *Phys. Metals Metallogr. (USSR)* **15**(4), 1 (1963) [*Fiz. Metal. Metalloved.* **15**(4), 481 (1963)].
176. K. Sadananda and M. J. Marcinkowski, *J. Appl. Phys.* **44,** 1989 (1973).
177. M. J. Marcinkowski and N. J. Olson, *Phil. Mag.* **19,** 1111 (1969).
178. M. J. Marcinkowski, N. J. Olson, and K. Sadananda, *Phys. Status Solidi (A)* **17,** 89 (1973).
179. K. Sadananda and M. J. Marcinkowski, *J. Appl. Phys.* **44,** 4445 (1973).
180. A. Seeger, *in* "Handbuch der Physik" (S. Flügge, ed.), Vol. VII/2, Kristallphysik II, p. 191. Springer-Verlag, Berlin and New York, 1958.
181. H. McI. Clark, *Phil. Mag.* **16,** 853 (1967).
182. A. Seeger, *in* "Dislocations and Mechanical Properties of Crystals" (J. C. Fisher, W. G. Johnston, R. Thomson, and T. Vreeland, Jr., eds.), p. 243. Wiley, New York, 1957.
183. M. J. Marcinkowski and H. J. Leamy, *J. Appl. Phys.* **40,** 3095 (1969).
184. R. W. Armstrong, *Advan. Mater. Res.* **4,** 101 (1970).
185. M. J. Marcinkowski and R. M. Fisher, *Trans. AIME* **233,** 293 (1965).
186. L. Y. Popov, E. V. Kozlov, and N. A. Aleksandrov, *Phys. Metals Metallogr. (USSR)* **21**(5), 108 (1966) [*Fiz. Metal. Metalloved.* **21**(5), 756 (1966)].
187. K. R. Jordan and N. S. Stoloff, *Trans. AIME* **245,** 2027 (1969).
188. K. R. Jordan and N. S. Stoloff, *Trans. Jap. Inst. Metals* **9,** Suppl., 281 (1968).
189. M. J. Marcinkowski, *J. Inst. Metals* **93,** 476 (1964–1965).
190. H. J. Leamy, P. Schwellinger, and H. Warlimont, *Acta Met.* **18,** 31 (1970).
191. G. E. Lakso, M.Sc. Thesis, Iowa State Univ., Ames, 1966.
192. P. Mouturat, J. Moinet, M. Romeggio, G. Sainfort, and G. Cobone, *J. Nucl. Mater.* **19,** 234 (1966).
193. D. L. Wood and J. H. Westbrook, *Trans. AIME* **224,** 1024 (1962).
194. F. X. Kayser, *WADC Tech. Rep., Ford Mot. Co.* **57-298,** Part I (1957).
195. M. J. Marcinkowski and R. W. Armstrong, *J. Appl. Phys.* **43,** 2548 (1972).
196. M. J. Marcinkowski and K. S. Sree Harsha, *J. Appl. Phys.* **39,** 6063 (1968).
197. F. Laves, *Naturwissenschaften* **30,** 546 (1952).
198. F. Laves, *Acta Met.* **14,** 58 (1966).
199. G. F. Bolling and R. H. Richman, *Acta Met.* **14,** 58 (1966).

200. R. W. Cahn and J. A. Coll, *Acta Met.* **9**, 138 (1961).
201. G. F. Bolling and R. H. Richman, *Acta Met.* **13**, 709 (1965).
202. G. F. Bolling and R. H. Richman, *Acta Met.* **13**, 723 (1965).
203. G. F. Bolling and R. H. Richman, *Acta Met.* **13**, 745 (1965).
204. R. H. Richman and G. P. Conard, II, *Trans. AIME* **227**, 779 (1963).
205. M. J. Marcinkowski and H. A. Lipsitt, *Acta Met.* **10**, 95 (1962).
206. N. M. Madhava, P. J. Worthington, and R. W. Armstrong, *Phil. Mag.* **25**, 519 (1972).
207. W. C. Leslie, D. W. Stevens, and M. Cohen, *in* "High-Strength Materials" (V. F. Zackay, ed.), p. 382. Wiley, New York, 1965.
208. B. N. Adrianovsky, V. I. Syutkina, O. D. Shashkov, and E. S. Yakovleva, *Phys. Metals Metallogr. (USSR)* **26**(5), 101 (1968) [*Fiz. Metal. Metalloved.* **26**(5), 874 (1968)].
209. V. S. Arunachalam and C. M. Sargent, *Scr. Met.* **5**, 949 (1971).
210. F. E. Wang, *J. Appl. Phys.* **43**, 92 (1972).
211. D. Hull, *Phil. Mag.* **7**, 537 (1962).
212. M. J. Marcinkowski, A. S. Sastri, and D. Koskimaki, *Phil. Mag.* **18**, 945 (1968).
213. A. S. Sastri, M. J. Marcinkowski, and D. Koskimaki, *Phys. Status Solidi* **25**, K67 (1968).
214. S. P. Gupta, A. A. Johnson, and K. Mukherjee, *Mater. Sci. Eng.* **11**, 29 (1973).
215. S. P. Gupta, A. A. Johnson, and K. Mukherjee, *Mater. Sci. Eng.* **11**, 43 (1973).
216. A. Nagasawa, *J. Phys. Soc. Jap.* **31**, 136 (1971).
217. F. E. Wang, S. J. Pickart, and H. A. Alperin, *J. Appl. Phys.* **43**, 97 (1972).
218. S. Rosen and J. A. Goebel, *Trans. AIME* **242**, 722 (1968).
219. A. Ball, *Metal Sci. J.* **1**, 47 (1967).
220. M. E. Brookes and R. W. Smith, *Metal Sci. J.* **2**, 181 (1968).
221. B. D. Masson and C. S. Barrett, *Trans. AIME* **212**, 260 (1968).
222. C. M. Wayman, *Scr. Met.* **5**, 489 (1971).
223. M. J. Marcinkowski and N. Brown, *Phil. Mag.* **8**, 891 (1963).
224. N. S. Stoloff, *in* "Fracture" (H. Liebowitz, ed.), Vol. 6, p. 1. Academic Press, New York, 1969.
225. N. S. Stoloff and I. L. Dillamore, *in* "Ordered Alloys—Structural Applications and Physical Metallurgy" (B. H. Kear, C. T. Sims, N. S. Stoloff, and J. H. Westbrook, eds.), p. 525. Claitor's Publ. Div., Baton Rouge, Louisiana, 1970.
226. J. F. Dinhut, J. P. Eymery, and P. Moine, *Phys. Status Solidi (A)* **12**, 153 (1972).
227. M. J. Marcinkowski, F. X. Kayser, and M. E. Taylor, Jr., *Met. Trans.* Submitted for publication (1974).
228. A. A. Presnyakov and G. V. Starikova, *Phys. Metals Metallogr. (USSR)* **12**(6), 84 (1961) [*Fiz. Metal. Metalloved.* **12**(6), 873 (1961)].
229. J. D. Meakin and N. J. Petch, *in* "Fracture of Solids" (D. C. Drucker and J. J. Gilman, eds.), p. 393. Wiley (Interscience), New York, 1963.
230. A. H. Cottrell, *Trans. AIME* **212**, 192 (1958).
231. M. J. Marcinkowski and J. Larsen, *Met. Trans.* **1**, 1034 (1970).
232. P. F. Timmins and A. S. Wronski, *Nature (London), Phys. Sci.* **235**, 113 (1972).
233. A. R. Bailey, R. McDonald, and L. E. Samuels, *J. Inst. Metals* **85**, 25 (1956–1957).
234. S. Harper, *J. Inst. Metals* **85**, 415 (1956–1957).
235. E. M. Grala, *in* "Mechanical Properties of Intermetallic Compounds" (J. H. Westbrook, ed.), p. 358. Wiley, New York, 1960.
236. A. V. Seybolt and J. H. Westbrook, *Acta Met.* **12**, 449 (1964).
237. J. H. Westbrook, *Met. Rev.* **9**, 415 (1965).
238. M. J. Marcinkowski, K. Sadananda, and W. F. Tseng, *Phys. Status Solidi (A)* **17**, 423 (1973).
239. M. J. Marcinkowski, *Phil. Mag.* **17**, 159 (1968).
240. J. Czernichow, J. Gudas, M. J. Marcinkowski, and W. F. Tseng, *Met. Trans.* **2**, 2185 (1971).

241. M. J. Marcinkowski and E. S. Dwarakadasa, *Phys. Status Solidi* (*A*) **19**, 597 (1973).
242. M. J. Marcinkowski, *in* "Electron Microscopy and Structure of Materials" (G. Thomas, R. M. Fulrath, and R. M. Fisher, eds.), p. 382. Univ. of California Press, Berkeley, California, 1972.
243. W. F. Tseng, Ph.D. Thesis, Univ. of Maryland, College Park, 1973.
244. D. R. Thornbury, *J. Appl. Phys.* **40**, 1579 (1969).
245. R. C. Boettner, N. S. Stoloff, and R. G. Davies, *Trans. AIME* **236**, 131 (1966).
246. R. C. Boettner and A. J. McEvily, Jr., *Acta Met.* **13**, 937 (1965).
247. J. R. Low, Jr. and A. M. Turkalo, *Acta Met.* **10**, 215 (1962).
248. T. Taoka and S. Sakata, *Acta Met.* **5**, 61 (1957).
249. T. Taoka and R. Honda, *J. Electronmicrosc.* **5**, 19 (1957).
250. V. I. Syutkina and E. S. Yakovleva, *Phys. Metals Metallogr.* (*USSR*) **14**(5), 92 (1962) [*Fiz. Metal. Metalloved.* **14**(5), 745 (1962)].
251. H. D. Williams and G. C. Smith, *Phil. Mag.* **13**, 835 (1966).
252. M. Herman and N. Brown, *Trans. AIME* **206**, 604 (1956).
253. N. Brown and D. R. Lenton, *Acta Met.* **17**, 669 (1969).
254. A. Gittins and R. C. Gifkins, *Acta Met.* **16**, 81 (1968).
255. R. G. Davies, *Trans. AIME* **227**, 277 (1963).
256. P. R. Strutt and R. A. Dodd, *in* "Ordered Alloys-Structural Applications and Physical Metallurgy" (B. H. Kear, C. T. Sims, N. S. Stoloff, and J. H. Westbrook, eds.), p. 475. Claitor's Publ. Div., Baton Rouge, Louisiana, 1970.
257. J. Bevk, R. A. Dodd, and P. R. Strutt, *Met. Trans.* **4**, 159 (1973).
258. A. Lawley, J. A. Coll, and R. W. Cahn, *Trans. AIME* **218**, 167 (1960).
259. A. J. R. Soler-Gomez and W. J. McG. Tagart, *Acta Met.* **12**, 961 (1964).
260. R. G. Davies, *Trans. AIME* **227**, 665 (1963).
261. J. B. Mitchell, *Scr. Met.* **4**, 411 (1970).
262. N. Brown, *Scr. Met.* **4**, 417 (1970).
263. A. H. Cottrell, "Theoretical Structural Metallurgy." Arnold, London, 1955.
264. J. M. Oblak and B. H. Kear, *in* "Electron Microscopy and Structure of Materials" (G. Thomas, R. M. Fulrath, and R. M. Fisher, eds.), p. 566. Univ. of California Press, Berkeley, California, 1972.

Subject Index

Page numbers in **bold type** denote the beginning of a chapter about the entry.

A 4
B 5
C 6
D 7
E 8
F 9
G 0
H 1
I 2
J 3